Sulfur Dioxide Sensors

Loveleen Kaur Gulati, Gurleen Kaur Gulati and Satish Kumar*

Department of Chemistry, St. Stephen's College, University Enclave,
Delhi 110007, India

*satish@ststephens.edu

Edited by

Inamuddin

Department of Applied Chemistry, Zakir Husain College of Engineering and Technology,
Faculty of Engineering and Technology, Aligarh Muslim University, Aligarh-202 002,
India

Published by **Materials Research Forum LLC**
Millersville, PA 17551, USA

Published as part of the book series
Materials Research Foundations
Volume 95 (2021)
ISSN 2471-8890 (Print)
ISSN 2471-8904 (Online)

Print ISBN 978-1-64490-122-9
ePDF ISBN 978-1-64490-123-6

Distributed worldwide by

Materials Research Forum LLC
105 Springdale Lane
Millersville, PA 17551
USA
http://www.mrforum.com

Printed in the United States of America
10 9 8 7 6 5 4 3 2 1

Table of Contents

The growing environmental pollution owing to the release of highly toxic sulfur dioxide (SO_2) gas is a cause of concern to the global community. The detection and quantification of sulfur dioxide in the environment is essential in order to prevent damage to our environment and improve the health of living organisms. A variety of sensing materials have been investigated and used for the detection of sulfur dioxide in the environment. The chapter focuses on the recent progress in sensing techniques, which are used for monitoring the presence of sulfur dioxide. A variety of sensors and biosensors have been discussed and compared in order to analyze the performance of different technologies, hence, providing the reader an overview of techniques best suited for the recognition of sulfur dioxide gas.

1.   Introduction

The energy usage by the growing human population and rapid industrial progress globally is leading to the emission of toxic gases into the atmosphere [1-3]. A variety of gases like CO, CO_2, NO_x, HCl, SO_x, NH_3 etc. are released into the atmosphere daily, which creates numerous environmental issues and affects the health of living organisms [1-4]. The environmental pollution is directly responsible for the ozone layer depletion, damage to structures, acid rain and global warming. Water, air and soil pollution represent three important categories of pollution that affect the human population around the world [2]. Among them, air pollution can have disastrous consequences for humans within a short span. In order to prevent damage to the environment due to pollution, it is essential to monitor the concentration of toxic gases or pollutants in the environment and biological systems. The air pollutants have been a subject of great interest for the scientific community to quantify and monitor pollutants in our environment. Among the toxic gases that pollute the environment, sulfur dioxide is a highly toxic, non-flammable, colorless gas with a characteristic pungent and suffocating odour. It is considered as a key air pollutant. The SO_2 emissions find its sources in both natural as well as anthropogenic processes. On a global scale [5], SO_2 emissions are predominantly due to natural sources, while, in urban and industrial regions, human activity plays a major role. Natural sources include volcanic eruptions [6]. Motor vehicle exhaust, combustion of fossil fuels, wood-burning stoves, smelters [7] etc. are the man-made sources that mainly contribute to SO_2 emissions. Sulfur dioxide gas is harmful once emitted into the atmosphere. This is because it reacts with various oxidants [8] present in the ambient air and is converted to sulfur trioxide (SO_3), bisulfite (HSO_3^-), sulfuric acid and other sulfates. Hydroxyl radical is accessed to be the prime oxidant that converts SO_2 to sulfate in daytime conditions (Scheme 1). Certain photochemical reactions in the troposphere

Materials Research Forum LLC

https://doi.org/10.21741/9781644901236

lead to its formation. Once formed, this radical oxidises SO_2 according to gas phase reactions (Scheme 1) [8].

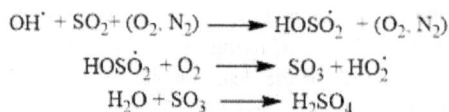

$$OH^{\cdot} + SO_2 + (O_2, N_2) \longrightarrow HOS\dot{O}_2 + (O_2, N_2)$$
$$HOS\dot{O}_2 + O_2 \longrightarrow SO_3 + HO_2^{\cdot}$$
$$H_2O + SO_3 \longrightarrow H_2SO_4$$

Scheme 1. *Formation of sulfate from SO_2 and $^{\cdot}OH$ [8].*

In humid environments, particulate sulfate formation is accelerated, as SO_2 (g) being water-soluble can be readily absorbed into the water droplets (Scheme 2) [8].

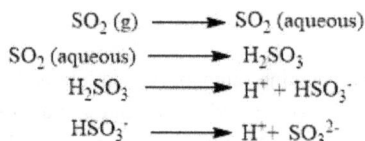

$$SO_2 \text{ (g)} \longrightarrow SO_2 \text{ (aqueous)}$$
$$SO_2 \text{ (aqueous)} \longrightarrow H_2SO_3$$
$$H_2SO_3 \longrightarrow H^+ + HSO_3^-$$
$$HSO_3^- \longrightarrow H^+ + SO_3^{2-}$$

Scheme 2. *Formation of sulfite ion from SO_2 (g) in the presence of water [8].*

The dissolved SO_2 is usually present in the HSO_3^-(aq) form in the pH range 3-6, which is the normal pH of atmospheric water droplets. The bisulfite ion formed in the atmospheric droplets undergoes aqueous oxidation by the two most probable oxidants O_3 and H_2O_2 to HSO_4^- in clouds and fog. These sulphates [9] [acidic sulfates such as H_2SO_4 and its neutralization products with atmospheric NH_3 e.g., letovicite $((NH_4)_3H(SO_4)_2)$, ammonium bisulfate (NH_4HSO_4), ammonium sulfate $((NH_4)_2SO_4)$] comprise the hazardous particulate matter, which remain suspended in the air and cause various environmental and health-related problems.

2. Environmental effects associated with SO_2 emissions

Sulfur dioxide in the atmosphere reacts with water vapors to form corrosive sulfuric acid, which is responsible for metal corrosion [5, 10]. It also damages building materials – limestone, marble, mortar, to name a few. High SO_2 concentration in the atmosphere lowers the soil pH [9] and, hence, can be detrimental to local vegetation. At low atmospheric concentration, when SO_2 enters a leaf, plant [6, 11] cells readily convert it to bisulfite and later to sulfate. But as its concentration in the environment increases, the rate of this conversion reaction drops and dilapidation of cell structure begins. Leafy vegetables including spinach, lettuce etc. are vulnerable to SO_2 exposure. The aerosols of sulfuric acid and other sulfates together formulate particulate matter, which is responsible for reduced visibility [6, 9, 12] due to its light-absorbing and scattering properties. This particulate phase also tends to reflect the incoming sunlight back into space and act as condensation nuclei. This causes the cloud reflectance to increase, thereby altering their lifetime, ultimately resulting in net cooling [11] of the earth's atmosphere.

3. Health effects of SO_2 exposure

Sulfur dioxide (SO_2) possesses high solubility in water. It enters the human body mainly through nasal airways [9] but little amount reaches the lungs. However, exposure to high concentrations of SO_2 for shorter durations multiple times can lead to irritation in the respiratory tract, shortness of breath, reflux cough, wheezing, chest tightness, etc. The biochemical mechanism [13] of SO_2 toxicity can be linked to its solubility in the aqueous phase. As mentioned earlier, SO_2 forms bisulfite in aqueous environments, thus reducing the pH slightly upon inhalation. In our body, the receptors for neurochemical reflexes are highly pH-sensitive and provide an immediate response upon pH change. As SO_2 is inhaled, a slight drop in pH generates the reflex broncho-constrictive response. Prolonged exposures to high atmospheric SO_2 levels may result in respiratory illness, worsening of cardiovascular diseases [6] (if any) and chronic lung diseases. In the asthmatic population and smokers, its effects are even more severe. The toxic effects of SO_2 on human health and surroundings urged the researchers to devise ways to monitor such harmful emissions. This chapter gives a summary of gas sensors that had been designed since the 1970s for sensing sulfur dioxide gas in particular, in actual atmospheric conditions.

4. Classification of gas sensors

During the last several decades, numerous studies have established different branches of gas sensing technologies [14, 15]. The major attention has been focused on the studies related to different types of sensors including the principle of sensing and their fabrication techniques. The gas sensors can be classified as electrochemical–potentiometric & amperometric; colorimetric and many others including semiconductor devices based on the method utilized for gas sensing. Tinoxide was widely studied for the detection of gaseous analytes. Fluorescence quenching is another useful and fascinating technique that was exploited for this purpose. Other sensors incorporated ionic liquids for electrochemical sensing, employed metallodendritic materials and adopted various unique methodologies like photoacoustic spectroscopy for detection of SO_2 in the gaseous phase. Biosensors area type of sensing devices that use biomolecules along with a transducer for the detection of sulfur dioxide. The evaluation of the gas sensing method is generally accomplished by considering parameters such as sensitivity, selectivity, response time, energy consumption, reversibility, absorption capacity, fabrication cost, stability of the sensing material and toxicity of chemical species used for sensing or the product formed during sensing process. In addition, if the sensing material is portable, it can be easily transported to remote areas for the desired application. A number of gas sensing materials have been developed for different types of toxic gases with high sensitivity, selectivity and stability. However, few have been reported to be reversible and produce less toxicity during the detection and quantification process. Due to increased awareness about environmental pollution, the sensing material for gas detection should not only detect the presence of a toxic gas molecule, but it should also be recoverable after the operation and produce less toxic side products.

The following section describes the different types of gas sensors in detail.

(I) Potentiometric gas sensors

Before detailing the research literature in this area, a brief understanding of potentiometric gas sensors is mandatory. Sensing through the potentiometric method [16] signifies that the chemical potential of the system varies due to the presence of the analyte, in the absence of current flow. Traditional potentiometric sensors [17, 18] for detection of gaseous analyte (also known as *Type I*) are a type of electrochemical sensors that consist of a solid electrolyte with reference and test compartments adjoined on either side of it. In this type, a solid electrolyte [19] contains the ions of the neutral gas component to be sensed. The conducting ions of the electrolyte establish equilibrium with the gaseous phase. The activity difference of these two components (in equilibrium) influences the potential difference existing between the two compartments (reference and test) across the solid electrolyte.

Usually, platinum metal is used as the electrode. The reaction occurring at the three-phase contact points (metal/electrolyte/gas) can be written as shown in scheme 3 [19].

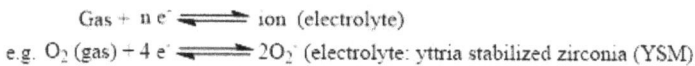

$$Gas + n\ e^- \rightleftharpoons ion\ (electrolyte)$$

$$e.g.\ O_2\ (gas) + 4\ e^- \rightleftharpoons 2O_2^-\ (electrolyte:\ yttria\ stabilized\ zirconia\ (YSM))$$

Scheme 3. *The reaction that takes place at electrode [19].*

The cell structure of the O_2 sensor [20]: O_2, $Pt^a|YSZ|^b$ O_2, Pt where a, b represents test & reference compartments, respectively.

According to the Nernst equation, emf generated from the cell is shown in equation 1 [19, 20].

$$E = \frac{RT}{4F}\ ln\ \frac{p^b}{p^a}\ \text{------------}\ \text{Equation 1}$$

Materials Research Forum LLC
https://doi.org/10.21741/9781644901236

where p^a, p^b represent oxygen partial pressures in test & reference compartments respectively. Here, the value of p^b should remain constant.

These *type I* sensors are meant for simple gases such as H_2, Cl_2, etc. and do not work for oxidic gases (CO_2, NO_2, SO_2) due to the unavailability of the corresponding solid electrolyte.

Type II sensors [17, 19, 20] are mainly functional for oxidic gases. Here, an analyte gas is in equilibrium with an electrolyte species that completely differs from the mobile component of the electrolyte. Salts of oxy-acids of the gas to be detected serve as the solid electrolyte.

E.g.: In the CO_2 sensor [21], K_2CO_3 works as a solid electrolyte.

Cell representation: CO_2-O_2, Au $^a|$ K_2CO_3 $|^b$ Au, CO_2-O_2

Here, K^+ions are the mobile species in the electrolyte. The electrodic reaction after interaction of CO_2 gas with the K_2CO_3 & Au electrode is shown in Scheme 4 [21].

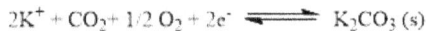

$$2K^+ + CO_2 + 1/2\ O_2 + 2e^- \rightleftharpoons K_2CO_3\ (s)$$

Scheme 4. *Electronic reaction [21].*

In this type of sensors, analyte gas is modified to form an immobile ion of the solid electrolyte.

Emf for the above cell reaction is shown in equation 2[21].

$$E = RT/2F\ \ln[(p_{CO2}^b \cdot p_{O2}^b)^{1/2} / (p_{CO2}^a \cdot p_{O2}^a)^{1/2}]\ \ldots\ldots\ \text{Equation 2.}$$

where p^a, p^b represents oxygen partial pressures in test & reference compartments respectively.

If $p_{O2}^a = p_{O2}^b$, then it simply becomes a CO_2 concentration cell, thus allowing the evaluation of p_{CO2}^b, if p_{CO2}^a is known.

SO_2 sensors broadly fall within the ambit of the Type II sensors.

In the case of *Type III* sensors [19, 22], an auxiliary phase is also present along with the solid electrolyte. Saito et al. [23] formulated the SO_2 sensor wherein a Na superionic conductor (NASICON) has been utilized as the solid electrolyte. The Na_2SO_4 was observed to be formed at its surface, thus served as the auxiliary phase. Cell configuration of this SO_2 sensor is as follows [23]:

SO_2-O_2-SO_3, Na_2SO_4, Pt $^a|$NASICON$|^b$ Pt, Na_2SO_4, SO_2-O_2-SO_3

Several SO_2 sensors come under this category.

Rosset al. [24] reported the development of a theoretical paradigm for studying the dependence of membrane characteristics on various parameters of gas detection (such as time of response, detection limit etc.). Several toxic gases were analysed but the study has been focussed on the analysis of SO_2 gas. The electrode for SO_2 recognition in the proposed model has been developed in parallelism with a Severinghaus [25] electrode. Bisulfite has been taken as the electrolyte and the reaction that occurred at the electrode is shown in Scheme 5 [24].

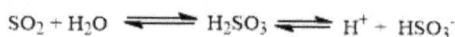

$$SO_2 + H_2O \rightleftharpoons H_2SO_3 \rightleftharpoons H^+ + HSO_3^-$$

Scheme 5. *The reaction that takes place at the electrode in the study [24].*

A pH electrode was employed to detect variations in the concentration of sulfur dioxide gas. Gas electrode design was as per Orion series 95 scheme (Fig. 1)[24].

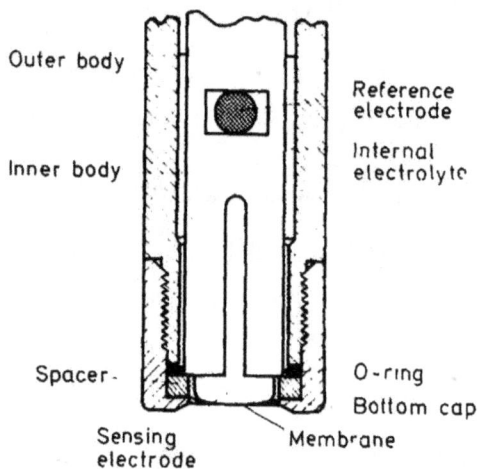

Fig. 1 *The layout of the Orion series 95 gas electrode. Reproduced here with the permission of the publisher IUPAC [24].*

The electrode response has been achieved at a minimum SO_2 concentration of 10^{-4} M.

As stated above in the introduction to potentiometric sensors, single-phase solid electrolytes were conventionally employed for gas detection. Worrell and Liu [26] have introduced a modification in the design through the employment of a two-phase solid electrolyte Li_2SO_4-Ag_2SO_4 in their neoteric SO_2 sensor.

In a later study, Pranitis et al. [27] have devised a sulfur dioxide sensing system utilizing a sulfite responsive membrane electrode based on neutral bis(diethyldithiocarbamato) mercury(II) complex. But the electrode was also responding to other background anions like iodide, bromide, thiocyanate, thiosulfate. To reduce such interferences, an electrode in combination with gas permeable membrane has been utilized, thus formulating the gas sensor selective for SO_2. The layout and its working principle have been demonstrated with the help of Fig. 2 [27].

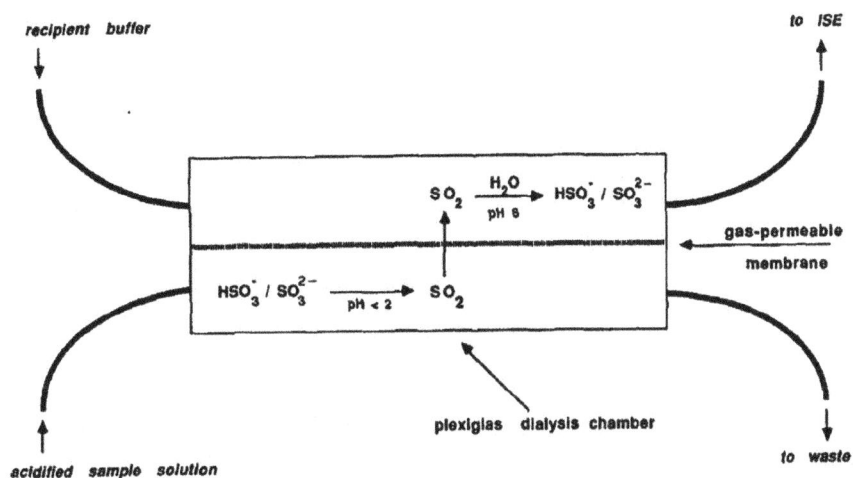

Fig. 2 *Diagrammatic representation of SO_2 sensor. Reproduced here with the permission of the Elsevier [27].*

The sample solution having HSO_3^-/SO_3^{2-} ions enters the chamber where these are converted to SO_2 at pH<2. The SO_2 gas generated through the process passes through the gas-permeable membrane to the other side, maintained at a pH = 6. At pH =6, the SO_2 gas is hydrolyzed back to HSO_3^-/SO_3^{2-}, which can be detected by an ion-selective electrode. The arrangement led to the achievement of high selectivity sensor for sulfite ions over iodide & bromide, though the interferences from thiocyanate and thiosulfate did not completely vanish [27].

Materials Research Forum LLC

https://doi.org/10.21741/9781644901236

The similar concept of placing the HSO_3^- selective electrode beyond the gas permeable membrane (to obtain sharp selectivity) was exploited by Mowery et al. [28]. The study involves multicyclic guanidiniumionophore (Fig. 3) [28] possessing a high affinity for oxoanions that formed the basis of the electrode membrane. The report demonstrated that this type of sensor (Fig. 4) [28] can be far more selective over the Severinghaus type SO_2 sensor with a detection limit of $2.8 \pm 1.2 \times 10^{-6}$ M (in terms of bisulfite).

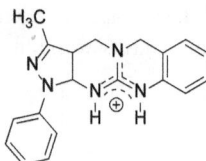

Fig. 3 *Guanidiniumionophore utilised in ion-selective electrode (ISE) [28].*

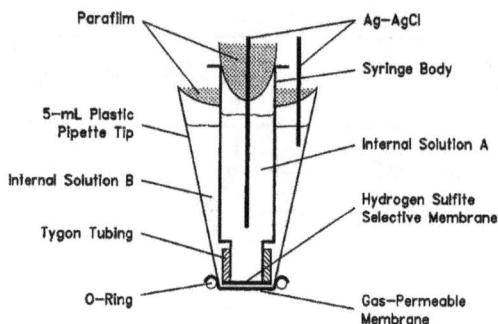

Fig. 4 *Sketch of SO$_2$ sensor depicting the placement of ISE second to the gas-permeable membrane. Reproduced here with the permission of ACS [28].*

The growth of potentiometric gas sensors has witnessed NASICON as the most preferred solid-state electrolyte [29] material due to its structural permanence [30, 31] and excellent conducting properties at relatively low temperatures. The productive range of composition of this structural class is the $Na_{1+x}Zr_2P_{3-x}Si_xO_{12}$ ($0 \leq x \leq 3$) solid solution, providing the highest ionic conductivity for $x=2$. In this structure, two salts namely $NaZr_2(PO_4)_3$ and $Na_4Zr_2(SiO_4)_3$, near the end show poor ionic conductivity [32]. Tiegang et al. [33] have developed a NASICON based SO_2 sensor incorporating a compound-metal oxide ($ZnSnO_3$) as the electrode material to receive an exceptional response in terms of selectivity and sensitivity towards low ppm levels of SO_2.Liang et al. [34] have utilized the same solid-state electrolyte but with a V_2O_5 doped TiO_2 electrode for efficient sensing of SO_2 gas in the range 1-50 ppm.

Shimizu et al. [35] have utilized a variant of NASICON in their sensor to improvise on its chemical stability and ionic conductivity at low temperatures. Sulfur dioxide (SO_2) detecting device comprising a sodium ionic conductor ($Na_5DySi_4O_{12}$), i.e. NaDyCON disk along with metal sulfide electrodes: metal monosulfides, disulfides & thiospinels (instead of stereotype metal sulfates as the auxiliary phase) were utilized.

Higher stability under acidic conditions and better ionic conductivity at comparatively low temperatures have been discovered, particularly for metal monosulfide $Pb_{1-x}Cd_xS$ (x=0.1,0.2) based sensor. Plots of EMF vs log SO_2 concentration were linear in the range 40 to 400 ppm depicting the enhanced potentiometric sensing.

In the development of gas sensors, perovskite oxide structures have emerged as convincing materials for sensing applications [36-38]. These are known for their electrocatalytic properties (beneficial to assist redox reactions occurring in an electrochemical sensor), high conductivity, and working stability at high operating temperatures, making these ideal to be exploited in gas sensors fabrication. Ma et al. [39] in a recent discovery, utilized the $La_xSm_{1-x}FeO_3$ electrode in conjunction with NASICON electrolyte to develop an SO_2 sensor thereby achieving sub-ppb level detection of SO_2. In this work, perovskite-kind orthoferrite $La_xSm_{1-x}FeO_3$ was studied for the effect of various doping proportions of lanthanum on the sensing properties. The highest response value (-86.5mV) was attained when the $La_{0.5}Sm_{0.5}FeO_3$ sensing electrode was subjected to 1.0 ppm SO_2 concentration (Fig. 5) [39]. Moreover, the sensor was able to detect as low as 5.0 ppb SO_2 concentrations with a response value of -8.4mV.

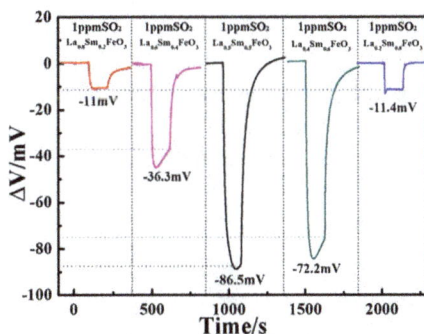

Fig. 5 *Response characteristics of the gas sensor linked to $La_xSm_{1-x}FeO_3$ as sensing electrode with x = 0.2, 0.4, 0.5, 0.6, 0.8. Reproduced here with the permission of Elsevier [39].*

SO$_2$ sensing mechanism was proposed based on literary articles on mixed potential theory [40-42]. The electrochemical cell formulated upon SO$_2$ exposure and the associated electrodic reactions [34] that took place at the electrode surface are shown in Scheme 6 [40-42].

$O_2 + SO_2$, La$_x$Sm$_{1-x}$FeO$_3$ -SE, Au| NASICON |Au, $O_2 + SO_2$

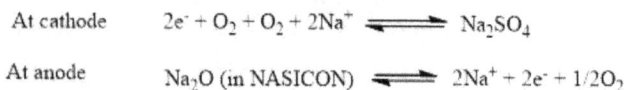

At cathode $2e^- + O_2 + O_2 + 2Na^+ \rightleftharpoons Na_2SO_4$

At anode Na_2O (in NASICON) $\rightleftharpoons 2Na^+ + 2e^- + 1/2O_2$

***Scheme 6**. The reaction taking place in the cell [40-42].*

In yet another mixed potential type SO$_2$ sensor, Liu et al. [43] have picked a novel columbite type composite oxide MnNb$_2$O$_6$ for sensing electrode material with zirconia (solid-state electrolyte) to detect low levels of SO$_2$ at high temperatures. The detection limit evaluated was as low as 50 ppb at 700°C exhibiting high-temperature stability and excellent selectivity.

(II) Amperometric gas sensors

In the amperometric mode of sensing [16, 44], the analyte concentration is examined through the faradic current generated due to the exchange of electrons between the working electrode (WE) and redox species (of the analyte) in the solution. The intensity of the current produced is proportional to the analyte concentration and thus gives rise to the sensor signal. In amperometric measurements, the applied potential is kept constant throughout the experiment. Amperometry involves the determination of both liquid phase as well as gas-phase analytes [45]. In the former case, electrodes and analytes are dipped in a common electrolyte solution but gaseous phase-detection requires a special gas (analyte) / liquid (electrolyte) / solid boundary (electrode) along with a transport system at the interface to manage the response characteristics of the gas sensor. Hence, fabrication of the sensor device should be such which allows gaseous analyte to effectively reach the electrode/electrolyte interface enabling it to participate in the redox reaction.

Basic layout of an amperometric gas sensor

Amperometric gas sensors are designed to incorporate three-electrode system viz. working electrode (WE), counter electrode (CE), and reference electrode (RE). At the working electrode, the potential is applied with respect to the reference electrode to initiate the electrochemical reaction. The counter electrode is usually employed to balance the reaction occurring at WE thereby facilitating the flow of current. All three electrodes are in contact with the electrolyte solution.

An essential feature typical of a gas sensor is the inlet route through which the sample gas is carried to the electrolyte/electrode interface, where the gas membrane allows selective diffusion of the analyte gas at the interface. The porosity of the electrode materials can be varied depending on the nature of the analyte gas to achieve selectivity in gas sensing. The elementary structure of a typical amperometric gas sensor is shown in Fig. 6 [45].

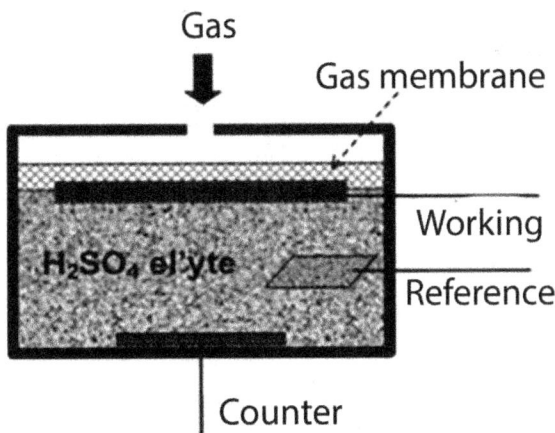

Fig. 6 *Demonstration of the amperometric gas sensor for CO with a three-electrode system. Reproduced here with the permission of ACS [45].*

Gas sensors based on the amperometric method of sensing have been abundantly applied in the field of environmental monitoring, and have registered vast improvement in the selectivity and sensitivity towards the detection of harmful pollutants. SO_2 gas sensors, in particular, are explored and discussed briefly hereunder [45].

Chiou et al. [46] have prepared a gold deposited electrode for gas diffusion in an amperometric SO_2 sensor, Fig. 7 demonstrating the overall sensor scheme. When the sensor was exposed to SO_2, its diffusion at the working electrode resulted in an oxidation reaction. A simultaneous reduction reaction commenced at the counter electrode (Scheme 7) [46].

At working electrode $\quad SO_2 + H_2O \rightleftharpoons SO_3 + 2H^+ + 2e^-$

At counter electrode $\quad 2H^+ + 2e^- \rightleftharpoons H_2$

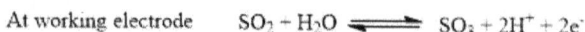

Scheme 7. The reaction occurring at different electrodes[46].

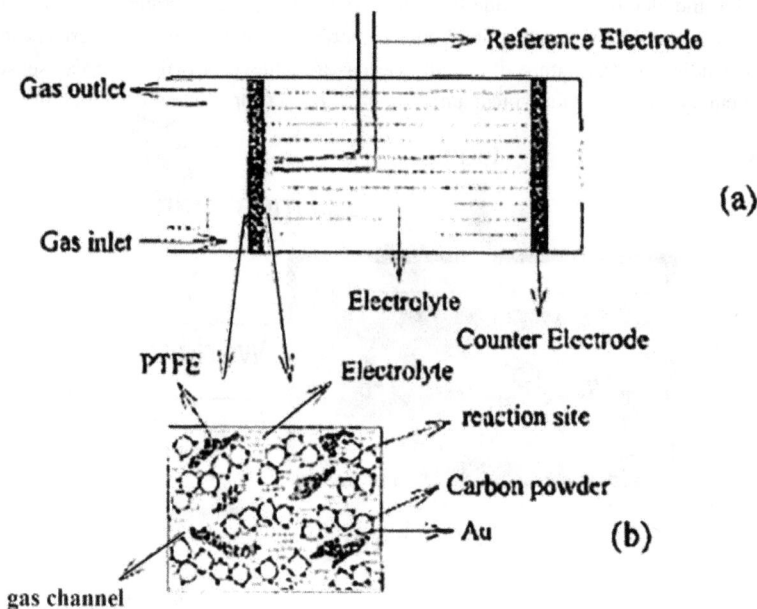

Fig. 7 The basic structure of a three-electrode mode gas sensor and the illustrative internal composition of the gas-diffusion electrode. Reproduced here with the permission of John Wiley and Sons [46].

The electric current produced between the two electrodes (at a constant applied potential of 0.6V) has been found to be proportional to the rate of oxidation of SO₂ at WE, enabling its quantitative detection. Fig. 8 [46] displays the proportionality of steady-state current vs SO₂ concentrations. The sensor device gave reproducible results manifesting high stability, quick response time and sensitivity of 1.002 $\mu Acm^{-2}ppm^{-1}$ of SO₂ in the range 25 to 500ppm.

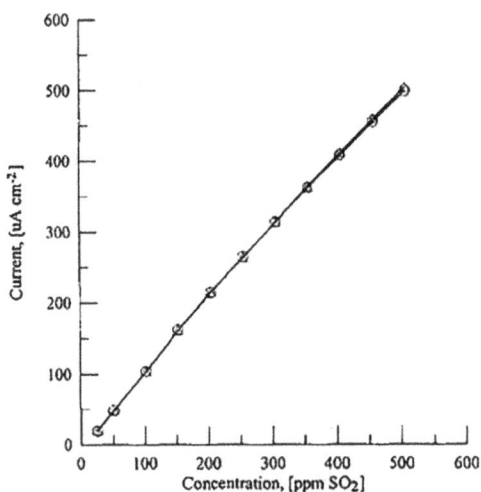

Fig. 8 *Steady-state current vs. SO₂ concentrations, at a constant potential of 0.6 V (vs. Ag/AgCl). Reproduced here with the permission of John Wiley and Sons [46].*

Shankaran and Narayanan [47] have reported an SO_2 sensor synthesis, wherein the graphite electrode using copper hexacyanoferrate (CuHCF) has been amended. Initially, the amperometric response of the sensor towards sulfite has been tested, and later it has been widened in approach to sense SO_2 gas in the detection range of 5.1 to 30.7 ppm. It has been envisaged that SO_2 should reach the sensor only after being absorbed through a sodium hydroxide solution having a pH between 3.5-8.0. During the analytical testing of the sensor, H_2S gas posed interference to the detection of SO_2. This was resolved with the help of Cu^{2+}ion solution that precipitated the sulfide, though up to 50 ppm H_2S concentration only.

Hodgson et al. [48] have announced a possibility of ppb level detection of atmospheric sulfur dioxide by making use of a noble metal Au - solid polymer electrode as the working electrode. Here, metal electrode [49] on one side experiences direct exposure to the gas phase, while the rear side of it is connected to an internal chamber housing an electrolyte solution with immersed reference and counter electrodes. Two types of solid polymer electrodes (SPEs)- nafion and ADP anion exchange membrane, have been employed in the study under contrasting pH conditions leading to three different cell setups:

(i) porous Au-nafion SPE in strong acid (internal electrolyte)

(ii) porous Au-nafion SPE in strong base (internal electrolyte)

(iii) ADP anion exchange membrane with the base as the internal electrolyte.

With all the three cell types, the intended low detection limits (in the ppb region) were realized. However, the use of alkaline internal electrolyte lowered the detection limit 3 times over the cell with acidic internal electrolyte. The application of the ADP anion exchange membrane as SPE did not show any demarcating results over nafion but was efficiently assembled to give reliable highly sensitive responses in SO_2 determination [48].

Voltammetry is also a type of amperometric technique in which current change is monitored as a function of dynamic applied potential [16]. Cyclic voltammetry [50] exhibits a huge potential in the area of gaseous analyte detection, though it is not as extensively exploited. It is an electrolytic technique that is used to monitor the oxidation-reduction processes of a molecular entity occurring during a chemical reaction. In a voltammogram (Fig. 9) [50], the applied potential is on the x-axis, while the resulting current is plotted on the y-axis. As the potential sweeps from high to low potentials (cathodic sweep), molecular species (say A^+) at the electrode gets reduced (leading to current generation), and its concentration in the vicinity of the electrode gradually decreases. As it continues, the peak current is attained. At this point, the current is generated due to the A^+ diffusion from the bulk solution. After this point, a steady decrease in current is observed because diffusion from bulk becomes difficult due to the formation of the depletion layer (reduced A^+). Current continues to fall till switching potential, where the direction of sweep is reversed and the reduced species (collected at the electrode) begins to oxidize back to A^+ (in an anodic sweep). In this way, the cycle is completed.

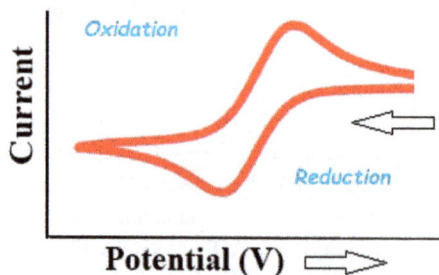

Fig. 9 *A typical cyclic voltammogram [50].*

Jasinski and Nowakowski [51] have exploited the above-mentioned technique for SO_2 gas detection with the help of a NASICON [52] type sensor. The technique involved the use of cyclic voltammetry for the determination of gas concentration. An 80 μm film of NASICON ($Na_{1+x}Zr_2Si_xP_{3-x}O_{12}$; x = 1.8) as solid electrolyte has been print on the electrode with a platinum electrode in the comb-type pattern (Fig. 10) [51]. Exposure of the sensor to SO_2 vapours produced characteristic current-voltage plots having positive and negative peaks (Fig. 11) [51]. Positive one signifying two distinct simultaneous processes- formation and decomposition of Na_2SO_4 at the two electrodes and negative peak depicting the reverse cycle (Fig. 11) [51], involving the following overall reaction (Scheme 8) [51].

$$Na_2SO_4 (s) \rightleftharpoons 2Na^+ + 2e^- + SO_2 + O_2$$

Scheme 8. *The decomposition reaction of sodium sulfate at electrode [51].*

SO_2 concentration in the atmosphere has been successfully determined using the technique employed in the study.

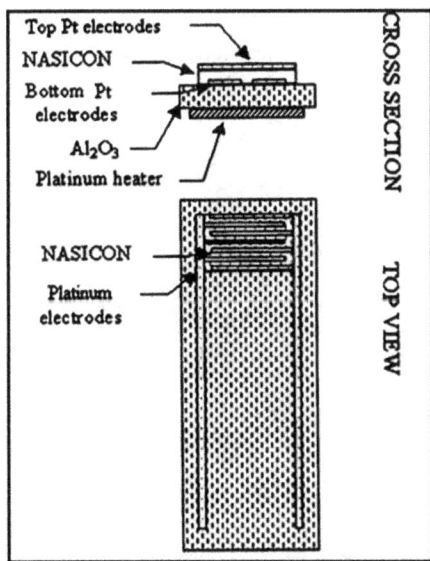

Fig. 10 *Sensor pattern. Reproduced here with the permission of Springer Nature [51].*

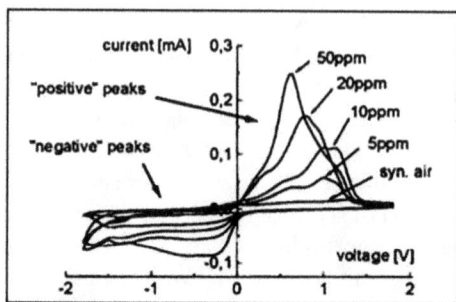

Fig. 11 *A plot of Current-voltage for different concentrations of SO₂. Reproduced here with the permission of Springer Nature [51].*

Jasinski et al. [53] have compared NASICON (with chemical formula-[$Na_{2.8}Zr_2Si_{1.8}P_{1.2}O_{12}$]) and LISICON [$Li_{14}Zn(GeO_4)_4$] based gas sensors for the detection of gases including SO_2 gas. This method also exploited cyclic voltammetry [54] for monitoring sensor response toward gases. On the CV (current-voltage) plots, LISICON has been found to show a higher peak current in comparison to the other; lithium-ion reactivity being the plausible cause. Reactions that led to peak generation in both the cases are shown in the schemes 9 and 10 [53].

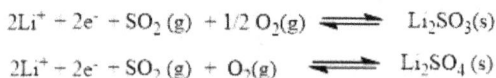

$$2Li^+ - 2e^- - SO_2\,(g) + 1/2\,O_2(g) \rightleftharpoons Li_2SO_3(s)$$

$$2Li^+ - 2e^- - SO_2\,(g) + O_2(g) \rightleftharpoons Li_2SO_4\,(s)$$

Scheme 9. *Formation of the lithium salt of sulfite and sulfate[53].*

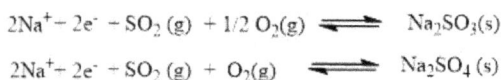

$$2Na^+ - 2e^- - SO_2\,(g) + 1/2\,O_2(g) \rightleftharpoons Na_2SO_3(s)$$

$$2Na^+ - 2e^- - SO_2\,(g) + O_2(g) \rightleftharpoons Na_2SO_4\,(s)$$

Scheme 10. *Formation of the sodium salt of sulfite and sulfate [53].*

Modifications in the working electrode materials have created immense possibilities for the development of unique and dependable SO_2 gas sensors [55]. Li et al. [56] have constructed an amperometric SO_2 sensor based on nano – Au assembled platinum working electrodes and utilized voltammetric mode for confirming the morphology of the electrode designed during the study. It has been further employed for investigation for sensing gaseous molecules. Sensing has been performed by varying internal electrolyte

16

solutions (HCl, NaOH, H_2SO_4, $HClO_4$) in order to arrive at cell conditions capable of providing maximum sensitivity and minimum response time. The 1.0 M NaOH solution was found to furnish the highest current responses to SO_2 at an applied potential of 0.6V and hence was concluded to be the most suitable internal electrolyte solution among others. Under these conditions, linearity was observed in the range 5 to 500 ppm SO_2 with a detection limit of 2.6 ppm. In addition to this, the selectivity of the sensor apropos sulfur dioxide was successfully established in the coexistence of other gases such as CO, NO, NH_3& CO_2.

Electrode surface refinement for enhancing the sensor response characteristics with reference to environmental monitoring has led the researchers to also consider self-assembled monolayer (SAM) techniques. SAM [57] based technology is a convenient way of constructing surfaces that can render immediate responses [58] to external stimuli. Electrochemical methods such as cyclic voltammetry are useful in understanding the reaction occurring at the monolayer surfaces. Shankaran et al. [59] have adopted this method for developing silver dispersed self-assembled electrode using (3-mercaptopropyl) trimethoxysilane (MPS) (Fig. 12) [59] in their SO_2 sensor. The objective was to add a catalytic effect of silver to the existing boons of self-assembled surfaces for the recognition of gaseous analytes at the electrode. This was demonstrated by performing the sensing experiments employing both glassycarbon(GC)/MPS/Ag electrode and bare electrode separately. Cyclic voltammograms of GC/MPS/Ag electrode (a) in the absence of sulfite indicated the redox reaction involving silver (anodic peak: +0.27 V, cathodic peak: +0.03 V) (Fig. 13) [59]. No such reaction was observed for the bare electrode. Further, on the addition of different sulfite concentrations (22.4, 60.8, 80 ppm) to the modified electrode, curves (b)-(d) were obtained against only one curve (f) that was noticed for bare electrode and that too for high SO_2 concentration (80 ppm).

Fig. 12 *SEM image of GC/MPS/Ag electrode. Reproduced here with the permission of Elsevier [59].*

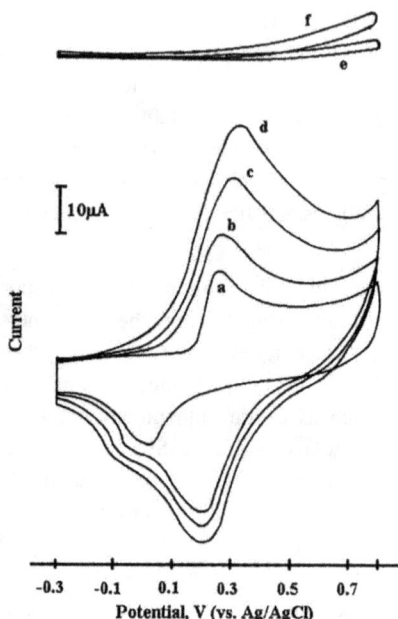

Fig. 13 *CV plots of GC/MPS/Ag electrode in 0.1 M KNO3 solution (a)showing absence of sulfite and presence of sulfite at concentrations (b) 22.4 ppm, (c) 60.8 ppm, and (d) 80 ppm. Curve (e) and (f) are the responses of bare GC electrode to the absence and presence of 80 ppm sulfite, respectively, pH: 7.0 (0.05 M phosphate buffer); scan rate: 50 mV/s. Reproduced here with the permission of Elsevier [59].*

The stability of the GC/MPS/Ag electrode has been investigated for 50 days (Fig. 14) [59], while stability under continuous hydrodynamic conditions has been investigated for 8 h (Inset of Fig. 14). No significant changes in response toward sulfite (30 ppm) have been observed during the first 10 days of intermittent storage of electrode in air and measurements (after every 2 or 3 days). The response reduced to 89% up to 30 days, which further reduced to 82% up to 50 days. The measurement of response under continuous stirring conditions was monitored for 8 h keeping the interval of response measurement at 1 h. The electrode has been stored in an electrolyte solution with continuous stirring throughout the experiment. The electrode response reduced to 91% under hydrodynamic conditions employed during the experiment (Fig. 14) [59].

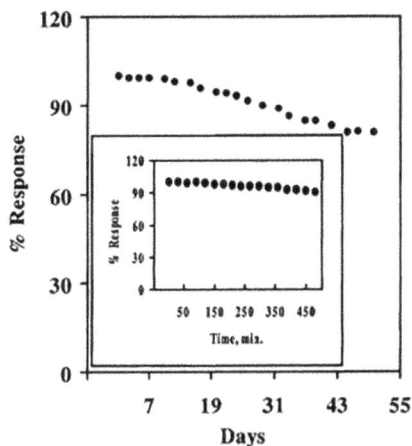

Fig. 14 *The response of the GC/MS/Ag electrode toward sulfite (30 ppm) sensing in 0.1M KNO₃ solution with time; pH 7.0; potential 0.3 V;stirring rate: 300 rpm; inset indicate sensing response under 8 h continuous stirring condition. Reproduced here with the permission of Elsevier [59].*

On switching from bare to the modified electrode, a substantial increase in the response current was recorded for the reduced value of over-potential, corroborating the catalytic effect of silver onto the self-assembled structure. Thus, the sensor displayed better response attributes in terms of stability, time of response and reproducibility, providing desirable sensitivity in the SO_2 concentration range of 0.6-88 ppm.

In the discussion up to now regarding the growth of electrochemical sulfur dioxide sensors, working electrode (based on SPE) modification techniques have been introduced in respective literary mentions. These methods [60] can be summarised here as chemical deposition, metal vapour deposition, electrochemical deposition or painting. Strzelczyk et al. [60] in a recent work, have made use of one of the two chemical deposition procedures viz. Takenaka-Torikai method to fill the pores of the nafion membrane with gold particles, thereby enabling the sensor to achieve a sensitivity coefficient of 0.0558 mA/ppm cm^2 towards SO_2.

(III) Optical sensors involving colorimetric and fluorescence changes

Optical sensing [61] refers to examining the alterations in light intensity in response to an external stimulus. The light changes may fall in the UV or visible region. For this reason, optical chemical sensors usually employ emission (fluorescence) or absorption techniques. Among other detection methods, colorimetric and fluorescent sensors have gathered considerable attention because of the simple, convenient, low-cost fabrication and low detection limits offered [62]. Thus, optical sensors present a propitious substitute to currently available other popular sensors.

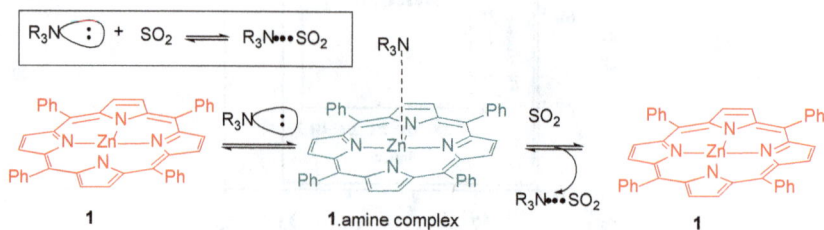

Fig. 15 *The SO₂ detection scheme in which indicator porphyrin **1** complex with amine followed by dissociation of the complex when exposed to SO₂ [63].*

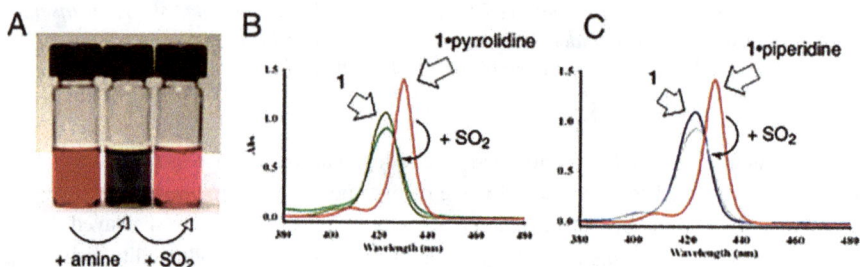

Fig. 16 *(A) Naked eye detection of SO₂: Leftmost vial-porphyrin **1**, then the addition of pyrrolidine, followed by SO₂. (B) Absorption spectra of porphyrin **1** and **1**•pyrrolidine complex, the addition of SO₂ resulted in a hypsochromic shift of the soret band arriving at initial λ_{max}= 421 nm corresponding to free porphyrin **1**. (C) Absorption spectra of porphyrin **1** and **1**•piperidine complex following the same procedure as (B). Reproduced here with permission from ACS [63].*

Leontiev et al. [63] have investigated a colorimetric SO₂ sensor exploiting donor-acceptor chemistry of amines, where SO₂ and a secondary or tertiary amine from a 1:1 complex. A variety of amines (pyrrolidine, piperidine, morpholine, diethylamine and quinuclidine) have been screened for affinity towards sulfur dioxide. The ¹H-NMR studies indicated a downfield shift in the α-CH signal of amine upon interaction with SO₂ gas. Therefore, as

Materials Research Forum LLC
https://doi.org/10.21741/9781644901236

per the sensing hypothesis (Fig. 15)[63] reported in the study, amine molecule forms a complex with metalloporphyrin and SO_2 is capable of displacing the porphyrin ring from complexed amine-metalloporphyrinmoiety, which ultimately results in a visual change in color (Fig. 16)[63]. The change in colour can also be monitored using UV-Visible spectroscopy (Fig. 16) [63]. Sulfur dioxide has also been selectively sensed in the presence of other exhaust gases like CO_x, NO_x and moisture, setting the detection limit at a low millimolar scale.

A very recent study in this domain also include amines, rather with a different perspective. Yu et al. [64] have enabled the colorimetric determination of SO_2 derivatives by a Co^{2+}ion prompted radical reaction pathway that eventually generates a highly conjugated photochromic entity. In this reaction, Co^{2+}ion gets oxidized to Co^{3+} by the action of dissolved oxygen, which in turn acts as an oxidant for the conversion of SO^{3-} or SO_2 to sulfate (Fig. 17) [64]. The sulfate finally oxidizes o-phenylenediamine (OPD) to OPD dimer and OPD trimer (capable of strong UV-vis absorption), giving rise to sensor signal (Fig. 18) [64].

The sensing system (CO^{2+} ion and OPD mixture) displayed a linear response to the increasing concentration of the SO_3^{2-} ions (Fig. 18) [64]. The sensing system displayed a good limit of detection value (4.07 μM) towards SO_3^{2-} ions (Fig. 18) [64].

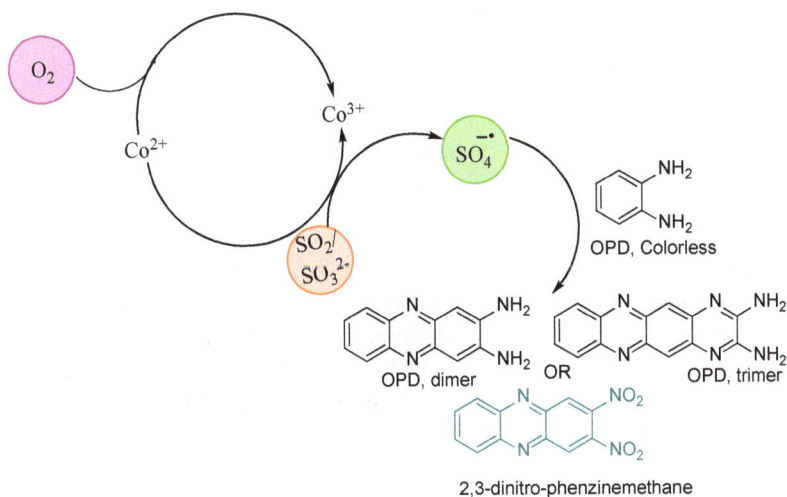

Fig. 17 *Schematic representation of the radical reaction set forth by Co^{2+} ions for colorimetric sensing of SO_2 derivatives [64].*

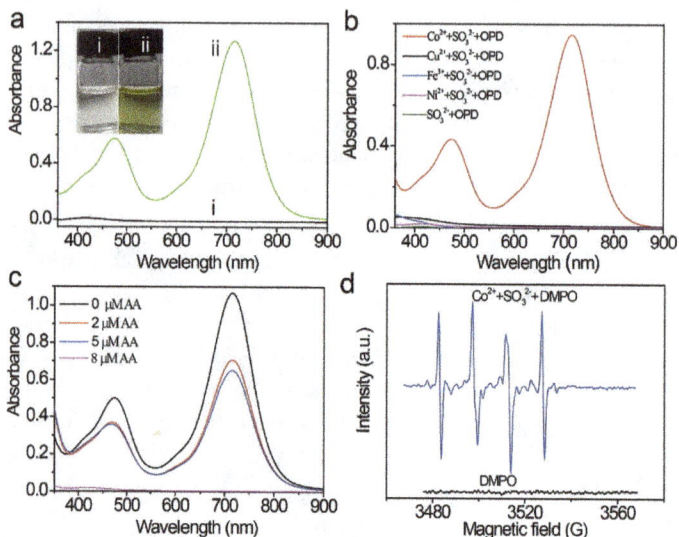

Fig. 18 *UV-Visible Spectrophotometric (Colorimetric) detection of SO_3^{2-} ion using Co^{2+} ion promoted radical reaction. (a) Changes in UV–Visible spectra for a Co^{2+} ion and OPD mixture upon addition of different concentrations of SO_3^{2-} (from bottom to top: 0, 10, 20, 30, 40, 50, 60, 150, 200, 250, 300, 350, 400, 450, 500, 550, 600, 650, 700, and 1000 μM). (b) A linear relationship between absorbance (at 716 nm) and the concentration of SO_3^{2-} ions. (c) A linear relationship between the change in absorbance at 716 nm and SO_3^{2-} ion concentration (60–650 μM). (d) Change in color in response to increasing concentration of SO_3^{2-} ions. Reproduced here with the permission of Elsevier [64].*

The bathochromic shift in the absorption band of the sensing system (Fig. 19a) [64] was explained through the computational calculation of the high occupied molecular orbitals (HOMO) and lowest occupied molecular orbitals (LUMO) orbitals of the OPD, OPD dimer and trimer. A decrease in the HOMO and LUMO gap has been indicated by computational calculations on the oxidative conversion of the OPD to OPD dimer and trimer (Fig. 19b) [64]. The colorimetric detection was discovered to exhibit linearity in the range 30-70 ppm of SO_2 with a detection limit of 1.26 ppm. Moreover, the system generated selective response to gases such as CO_2, NO_2, NH_3, O_2 & N_2 (Fig. 19) [64].

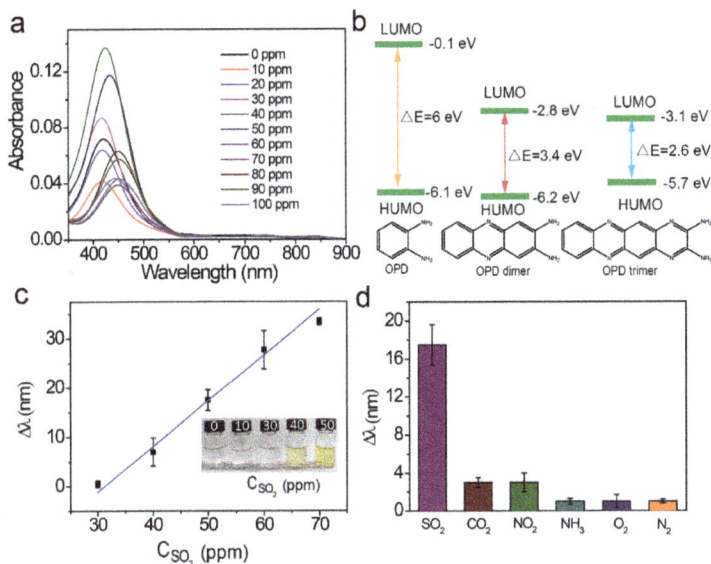

Fig. 19 *The analytical response of the sensing system for SO₂. (a) UV–Visible absorption spectra of the sensing system upon addition of different concentrations of SO₂ (0, 10, 20, 30, 40, 50, 60, 70, 80, 90, and 100 ppm). (b) The HOMO and LUMO energy states calculated for the OPD, OPD dimer, and OPD trimer. (c) The linearity between absorption wavelength change (Δλ) and SO₂ concentration in the range of 30–70ppm. The inset displays a snapshot of the Co²⁺-OPD based sensing system for the increasing concentration of SO₂ gas. (d) Sensor responses to other gases. Reproduced here with the permission of Elsevier [64].*

Now, switching on to fluorescence-based optical sensors, firstly fiber-optic technology is briefly described, which has been widely applied in the construction of these sensors.

An optical fiber [65] is a glassy thin fiber for light transmission. It comprises two layers-inner core having a higher refractive index and an outer coating with a lower refractive index. Under these circumstances, light is transmitted via total internal reflection between the two layers (Fig. 20) [65].

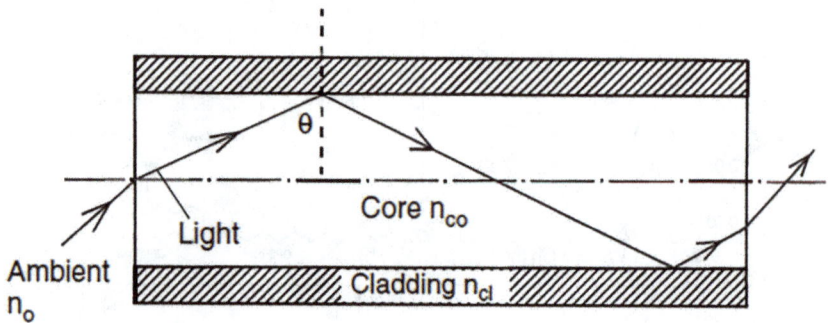

Fig. 20 *Showing light transmission through an optical fiber undergoing total internal reflection. Reproduced here with the permission of John Wiley and Sons [65].*

An important thing to be noted here is that, for total reflection to take place, a fraction of light energy infiltrates the cladding layer, and is regarded as evanescence field. A variation in this evanescence field with respect to an external disturbance is exploited for fiber-optic sensing applications [65]. These optical fibers furnish high sensitivities and remote sensing ability to the sensors. Wolfbeis et al. [66] have devised a fiber-optic sensor that is based on dynamic fluorescence quenching. Dynamic quenching [67, 68] occurs when a quencher diffuses to an excited state fluorophore, while static quenching results from the formation of a non-fluorescent ground state complex between fluorophore and quencher. The Stern-Volmer equation explains the correlation between fluorescence intensity and the concentration of the quencher (Equation 3).

$$\frac{F_0}{F} = 1 + K_{SV}[Q] \qquad \ldots\ldots \text{ equation 3}$$

where F_0, F represent the intensity of fluorescence in the absence & presence of quencher

$[Q]$ refers to the concentration of quencher

K_{SV} refers to stern-volmer constant.

Linear Stern-Volmer plot does not guarantee dynamic quenching, it is the fluorescence decay time, which is used to differentiate between the dynamic and static quenching of fluorescence [69]. Wolfbeis et al. [66] have utilized dynamic fluorescent quenching of polyaromatic hydrocarbons (PAH) agglutinated in a silicone polymer, by sulfur dioxide gas (acting as a quencher) as the basis of their SO_2 sensor device. The limit of detection has been estimated at 0.01% (v/v) SO_2 in ambient air, 0.01-6% (v/v) being the detectable

Materials Research Forum LLC
https://doi.org/10.21741/9781644901236

range. Atmospheric oxygen being an efficient quencher presented an obvious hindrance to the selective detection of the analyte gas, but below 6%(v/v) SO_2 levels in the air at constant O_2 pressures, its effect was quite insignificant (due to 26 folds greater quenching efficiency of SO_2 than O_2).

Razek et al. [70] have undertaken a similar approach for fluorescence quenching of membranes constituting rhodamine B isothiocyanate and silicone. Sensing could be performed for $0.114 \pm 0.009\%$ SO_2 as the lower limit. The sensor was selective for SO_2 amidst the rest of the atmospheric gases (HCl, NH_3, NO, CO_2) except O_2 which altered the sensor response.

Farooq et al. [71] have utilized the same fluorophore (Rhodamine B isothiocyanate-RBITC) after immobilizing it within silica nanoparticles to develop an SO_2 nano-sensor. Fig. 21 [71] depicts the device used for fluorescence measurements.

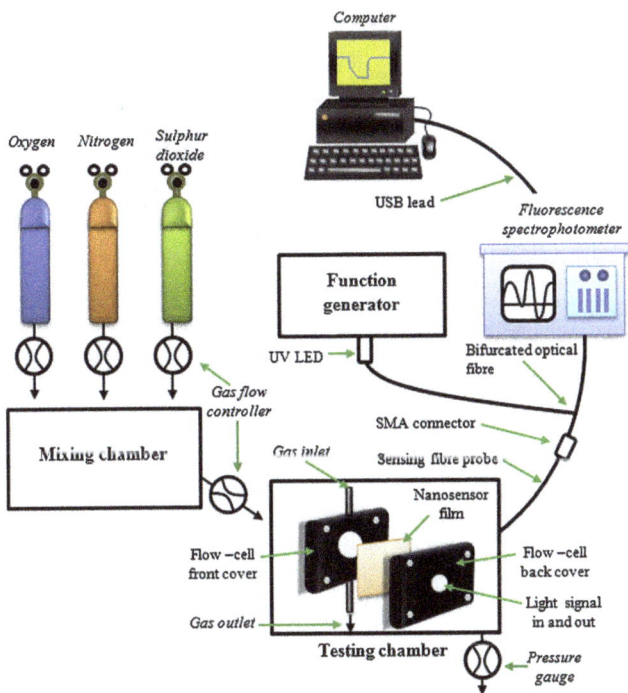

Fig. 21 *Diagrammatic representation of fluorescence determining instrument. Reproduced here with the permission of Elsevier [71].*

Materials Research Forum LLC

https://doi.org/10.21741/9781644901236

Sulfur dioxide was able to achieve a considerable amount of quenching of the fluorescence intensities, which helped in attaining a detection limit at 50%. Fig. 22 [71] & 23 [71] show initial fluorescence of RBITC and its consequent quenching when laid open to different concentrations of SO_2 in the N_2 atmosphere, respectively.

A

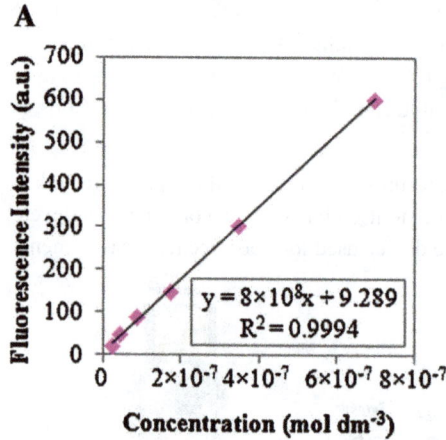

Fig. 22 Calibration plot of fluorescence of RBITC. Reproduced here with the permission of Elsevier [71].

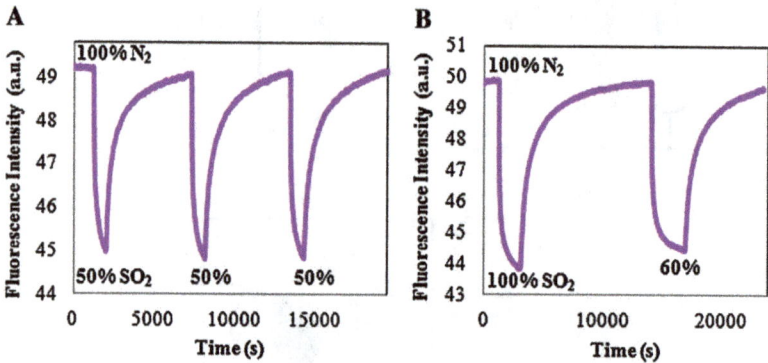

Fig. 23 Quenching of fluorescence RBITC functionalized SiNPs when subjected to (A) cyclic concentration of SO_2 & (B) 100% and 60% SO_2 under N_2 atmosphere. Reproduced here with the permission of Elsevier [71].

Apart from quenching, fluorescence enhancement is also a useful tool that is practiced in the fabrication of gas sensors. Sun et al. [72] have synthesized a turn-on fluorescent sensor employing carbon dots functionalized with a layer of cyanine dye. The proposed mechanism (Fig. 24) [72] was based on fluorescence resonance energy transfer (FRET) process. Initially, cyanine coated carbon dots were prepared, wherein the absorption spectrum of cyanine overlapped with the emission spectra of carbon dots, resulting in a very weak green fluorescence due to FRET. On the addition of bisulfite to the cyanine-carbon dots solution, bisulfite managed a nucleophilic attack at an α, β-unsaturated bond in the dye, thereby blocking the energy transfer pathway. This facilitated the restoration of the fluorescence to bright blue under a UV-lamp, also visible to the naked eye.

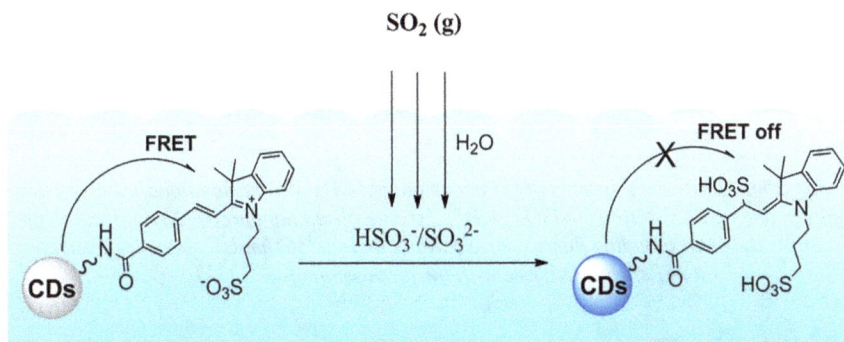

Fig. 24 *Outlines the SO₂ detection strategy. Reproduced here with the permission of ACS [72].*

Fig. 25 [72] displays the fluorescence intensities at subsequent steps in SO$_2$ detection. The sensor could efficiently sense as low as 1.8 μM bisulfite with a correlation coefficient (R^2) as 0.9987.

Li et al. [73] provided a sensitive fluorescence detection of SO$_2$ by engineering the surface chemistry of 3-aminopropyltriethoxysilane (APTES) modified quantum dots deploying a fluorescent coumarin-3-carboxylic acid (CCA) sensitive to SO$_2$. As per the detection scheme, the carboxyl group in blue emissive CCA reacted with the amino group of APTS, forming a charge-transfer complex. This caused a reduction in the fluorescence emission intensity of CCA in the complex due to protonation by amino groups. On exposure to SO$_2$, CCA was displaced from the complex and a new 1:1 charge-transfer complex was formed between SO$_2$ and the amino groups of APTES. This refurnished the blue fluorescence emission characteristic of free CCA molecular structures. Fig. 26 [73] elucidates the detection mechanism of SO_2. The sensor has been proved useful in detecting gaseous SO_2 at a concentration not below 6.0 ppb, amongst other co-present gases.

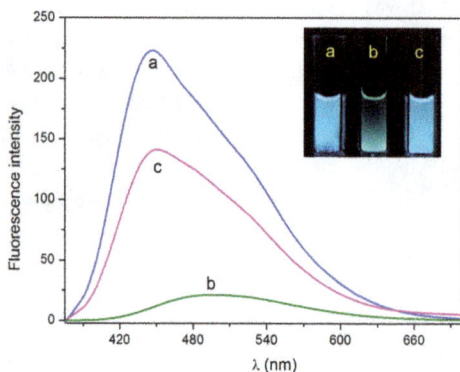

Fig. 25 *Fluorescence spectra of (a) amine coated CDs, (b) the functionalized Cy-CDs nanoprobe, and (c) action of HSO₃⁻ on the functionalized nanoprobe. The infixed picture shows the corresponding fluorescence colours under a 365nm UV lamp, respectively. Reproduced here with the permission of ACS [72].*

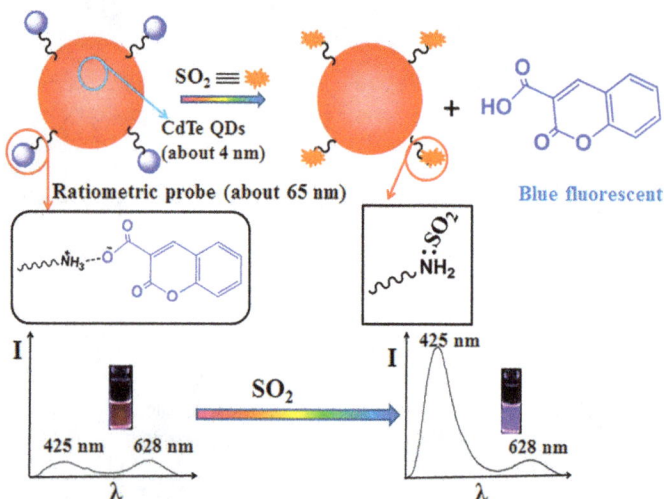

Fig. 26 *Sketch describing the principle of visual identification of SO₂. Reproduced here with the permission of ACS [73].*

Another recent study by You et al. [74] have described the synthesis of hydrophobic silica aerogels housing the fluorescent perovskite quantum dots (PODs), whose fluorescence can be selectively, reversibly and efficiently quenched by SO_2 (Fig. 27) [74].

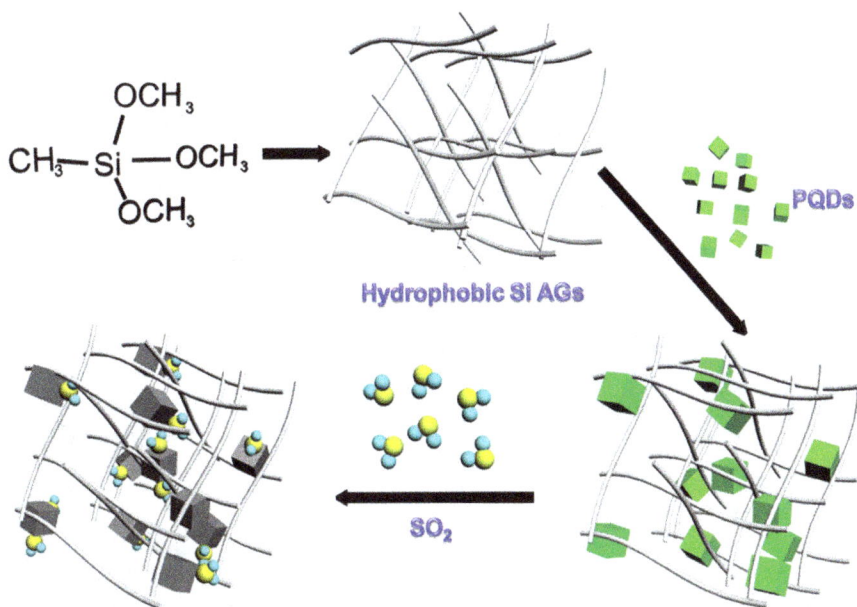

Fig. 27 *Demonstrating the principle of SO_2 detection. Reproduced here with the permission of ACS [74].*

Another recent research concerning optical sensors is based on an amino-functionalized luminescent metal-organic framework with a turn-on ability of detecting SO_2. "Metal-organic framework" (MOF) [75, 76] term has been coined for networked systems with sizeable and permanent porosity, formed with the conglomeration of metal centers and organic bridging ligands. This type of structure encompasses high architectural tenability and rich surface chemistry. MOF has thus emerged as crucial porous materials overcoming the shortcomings of hitherto porous solids including zeolites, mesoporous silica, activated carbon etc. Owing to the advantageous features, MOFs have found applications in various fields such as gas adsorption, gas separation and storage, sensing

[77, 78] and catalysis. Wang et al. [79] in the aforementioned work constructed an amino group-based MOF-5 (Fig. 28) [79] for SO_2 sensing applications.

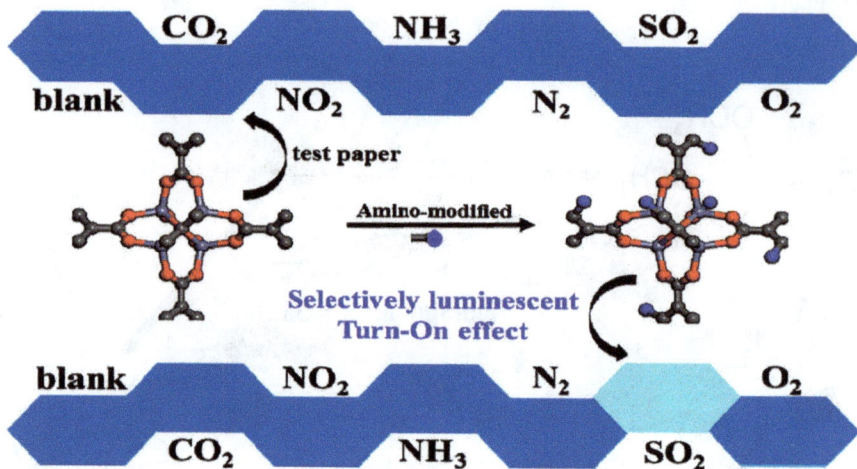

Fig. 28 *Indicating the role of amino-group in SO_2 identification. Reproduced here with the permission of ACS [79].*

The basic idea behind luminescence was the charge transfer mechanism that was supposed to occur on the interaction of SO_2/SO_3^{2-} with $-NH_2$ groups attached to MOF-5. The sensor can be considered for practical utilization with a quick response time of 15s and a lower limit of detection value (0.168 ppm). Fig. 29 [79] displays the luminescence intensities of MOF with & without $-NH_2$ functionalization asserting the importance of $-NH_2$ in the detection of SO_2. The detection can also be performed on a paper strip (Fig. 29c) [79].

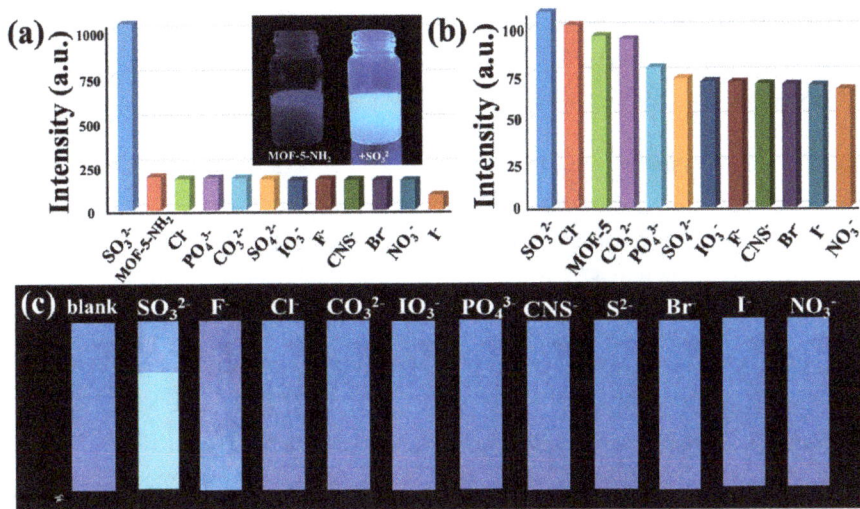

Fig. 29 *(a) A plot of luminescence intensities of MOF-5-NH₂ vs different anions at an identical concentration (2 × 10⁻⁴ M). The inset shows photoluminescence of MOF-5-NH₂ without and with SO₃²⁻, respectively. (b) A plot of luminescence intensities of un-modified MOF-5 interacting with different anions at an identical concentration (2 × 10⁻⁴ M). (c) A picture showing the detection of sulfite ion on a paper strip under UV-light of 365 nm. Reproduced here with the permission of ACS [79].*

(IV) Ionic liquids (ILs) based sensors

"Ionic liquids (ILs)" is a term that describes two properties of a chemical composition owing to its nomenclature, i.e., ionic and liquid; Ionic-describes that it is a salt comprising of cations and anions from different origins, e.g., organic or inorganic; and liquid-tells that ILs are liquid below 100°C. An example of room temperature ionic liquid (RTIL) is ethylammoniumnitrate (EAN) which has a melting point of 12°C [80].

In literature, ILs are better described as "molten salts". These are different from normal salts in that such molecules contain two oppositely charged asymmetrical ions not so tightly held to each other (due to incompatible size). This loose arrangement of the ions is the cause of these being liquid at room temperature [81].

ILs possess a large variety of properties due to which it has become an attention seeker for many researchers. These are thermally and chemically very stable, least volatile, non-flammable, have high ionic conductivity, low toxicity, etc. Moreover, its high pliability in molecular structure designing is also exploited for various applications. The properties exhibited by ILs are a function of the nature of cation and anion, which can be varied depending on the application [82].

High viscosity (suitable for binding the molecule to the electrode), broad electrochemical window, low current densities [83], high ionic conductivity, low vapour pressure are some of the important properties of ILs for these to be used in electrochemical devices [81, 84].

Based on solubility considerations, ILs can be classified as Hydrophobic ILs and Hydrophilic ILs. The former type is suitable for electrochemical systems as they form interface, which can't be polarized on coming in contact with an aqueous medium. An example of hydrophobic IL is [BMIM][PF_6] i.e. 1-butyl-3-methylimidazolium hexafluorophosphate. [BMIM][BF_4] is an example of hydrophilic IL, which are found to be unstable in aqueous media [81].

The simple chemical sensors are essentially judged on the basis of the type of components and their organization, of which electrolyte is a fairly important constituent that decides their efficiency and shelf-life, to a large extent. Conventional sensors based on organic electrolytes (having high volatility) are lacking in the sense that their vaporization takes place with time. ILs based electrochemical sensors, on the other hand, employ ILs as the electrolytes [85-87] in the electrochemical assembly. RTILs are least volatile with high chemical stability. This property makes them perfectly suitable to be served as an electrolyte media in robust gas sensors. Moreover, the presence of RTILs as electrolytes excludes the use of supporting electrolytes due to their high conductivity. RTILs also provide wide potential windows, which proves to be useful for studying the compounds that may have been inaccessible otherwise. RTILs solubility in gases adds to the advantages of them being used as electrolytes for the detection of gases [88]. Recent discoveries on RTILs (in amperometric gas sensor) [89] do away with the need of the membrane of the electrodes, which may improvise the response time of the detection of analyte gas.

In other formats, ILs are used to modify the electrodes by acting in combination with other additives.

Mu et al. [90] have developed a miniaturized SO_2 gas sensor with an electrochemical approach that utilized RTILs. The RTILs coated metal electrodes have been utilized having a porous polytetrafluoroethylene (PTFE) membrane on the other side by introducing a planar-electrodes-on-permeable membrane (PEoPM) gas sensor structure. The presence of permeable membrane [85] enables faster diffusion of the gases to reach the electrodes/RTIL interface. The properties of RTILs such as nominal vapor pressure, high viscosity, broad potential windows, and high thermal stability made these suitable to be exploited in an electrochemical sensor. The 1-butyl-1-methylpyrrolidinium bis(trifluoromethylsulfonyl)imide, [C$_4$mpy][NTf$_2$] has been employed as the RTIL in the experiment. Sensing property for the detection of toxic gas SO$_2$ and explosive gas CH$_4$ were investigated. Since the permissible exposure limit of SO$_2$ is 5 ppm, hence, SO$_2$ concentrations in the range 0 to 5 ppm were used to obtain a calibration curve through the measurement of impedance amplitude values at 1.0 Hz. Through this study, a compact RTIL based gas sensor with a PTFE membrane structure has been developed that can serve as a multi-gas sensing microsystem to monitor human health and safety [90].

Lichan Chen [91] have developed an IL-based sensor that works by quenching of electrochemiluminescence (ECL) of excited state luminophore, Ru(bpy)$_3^{2+}$/O$_2$(present in the IL film) in response to SO$_2$ molecules even at ppb level concentration. The mechanism can be explained through a diagrammatic scheme shown in Fig. 30 [91]. However, the sensor cannot be used for practical sensing of SO$_2$ in air samples due to coexisting molecular oxygen. The oxygen acts as a quencher at high concentration.

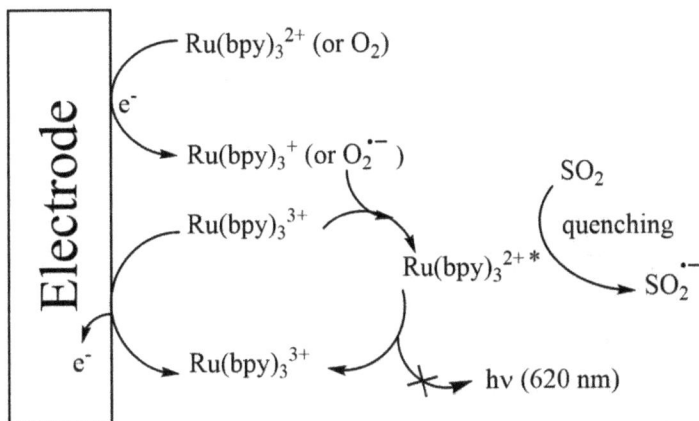

Fig. 30 *SO$_2$ effecting quenching for Ru(bpy)$_3^{3+}$/Ru(bpy)$_3^+$ & Ru(bpy)$_3^{2+}$/O$_2$ ECL systems* *[91].*

Chi et al. [92] have synthesized an anion-functionalized ionic liquid [P$_{66614}$][HBO] (Fig. 31) [92] exhibiting blue fluorescence, which was later quenched by the action of SO$_2$.

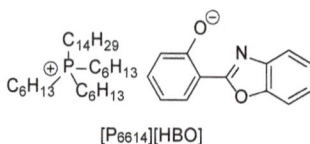

Fig. 31 *Structure of ionic liquid [P$_{66614}$][HBO] [92].*

ILs were chosen as they show high gas absorption capability, good selectivity and outstanding reversibility - properties detrimental for gas detection. Fig. 32 [92] signifies the fluorescence emission of the designed IL (at λ_{ex}=400 nm) and its subsequent suppression when exposed to SO$_2$. In addition, the sensor delivered reproducible responses (over 5 cycles) without any diminution observed in the initial performance (Fig. 33) [92].

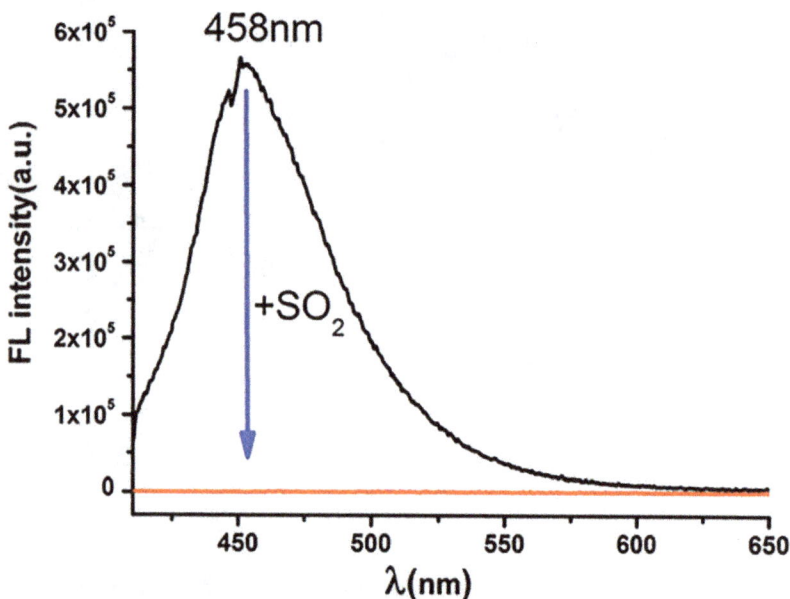

Fig. 32 *Fluorescence emission spectra of pure ionic liquid [P$_{66614}$][HBO] in the presence and absence of SO$_2$ (λ_{ex}=400nm by UV lamp). Reproduced here with the permission of RSC [92].*

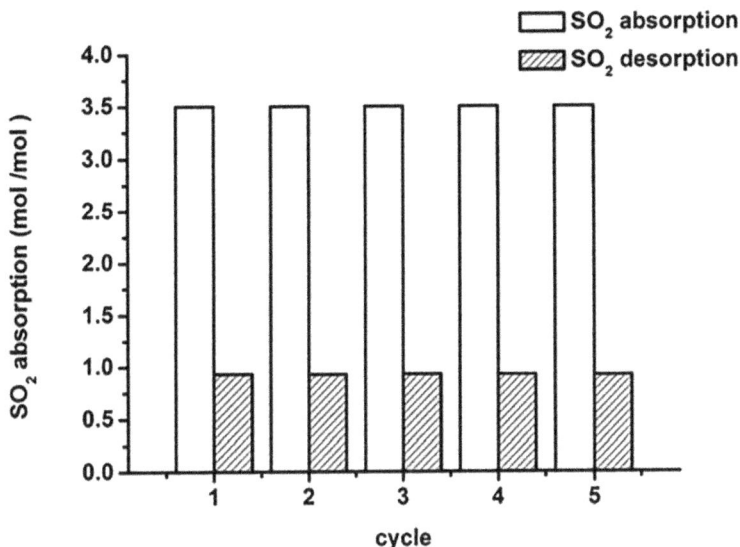

Fig. 33 *The reproducible character of the sensor to detect SO₂: adsorption at 30°C & 1 bar under SO₂ (60 ml min⁻¹) for 15 min., desorption at 90°C & 1 bar under N₂ (100 ml min⁻¹) for 60 min. Reproduced here with the permission of RSC [92].*

Xu et al. [93] in a recent study have presented a novel approach for SO_2 detection involving ILs with an interplay of electrochemical sensing techniques. The study utilized two ILs: [EMIM][TfO] and [EMIM][BF₄] (differing in the type of anion) for the purpose of adsorption and desorption of SO_2, to be detected at MoS_2/nafion modified glassy carbon electrode (Fig. 34) [93].

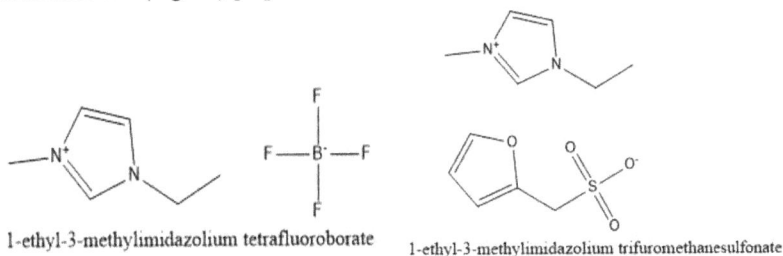

1-ethyl-3-methylimidazolium tetrafluoroborate

1-ethyl-3-methylimidazolium trifuromethanesulfonate

Fig. 34 *Structures of the ionic liquids: [EMIM][BF₄] and [EMIM][TfO] [93].*

ILs have shown a high SO_2 absorption capacity. The MoS_2-modified electrode has also been judged to be a good SO_2 sensor. On comparison, [EMIM][TfO] has been found to be superior of the two ILs in terms of absorption capacity, the probable reason being the higher solubility of SO_2 in it, maybe due to molar volume considerations ([BF$_4$]<[TfO]). Cyclic voltammograms of [EMIM][TfO] and [EMIM][BF$_4$] post SO_2 absorption are given in Fig. 35 [93].

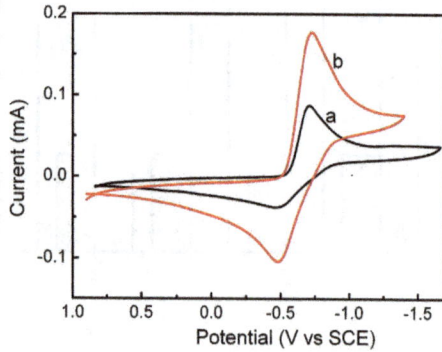

Fig. 35 *CV of MoS$_2$-modified electrode depicting 0.31% SO$_2$absorption in ILs (a)[EMIM][TfO] and (b) [EMIM][BF$_4$]. Reproduced here with the permission of Elsevier [93].*

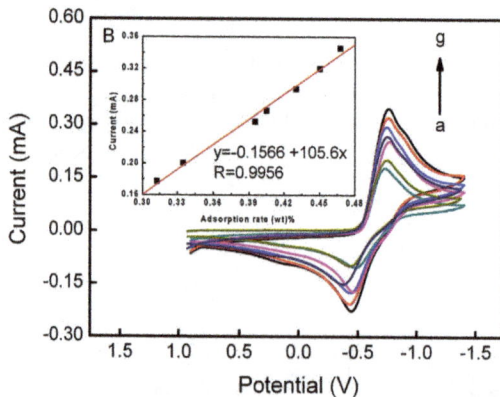

Fig. 36 *CVs of [EMIM][TfO] with varied amounts of absorbed SO$_2$, a-g symbols denote absorption rates-0.31%, 0.33%, 0.39%, 0.41%, 0.43%, 0.45%, and 0.47%, respectively. Inset: graph of peak current versus absorption rate. Reproduced here with the permission of Elsevier [93].*

Fig. 36 [93] depicts SO_2 sensing using [EMIM][TfO] at varied absorption rates. High response characteristics allowed the exploitation of the sensor to detect the gas in real-time surroundings and in the hazy atmosphere. In addition, the sensor displays a linear response toward SO_2 concentration in air (Fig. 37) [93].

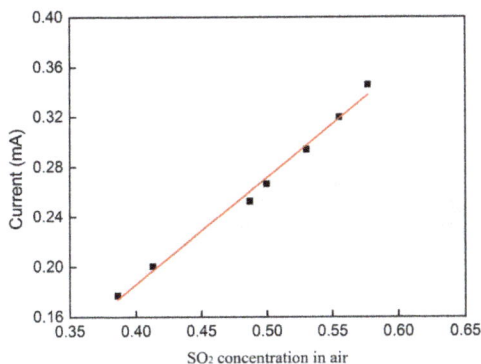

Fig. 37 *A linear relationship between sensor response current and the concentration of SO_2 in air. Reproduced here with the permission of Elsevier [93].*

(V) Semiconducting metal-oxide based sensors (SMOXs)

Metal-oxides can be employed for the detection of combustible, reducing or oxidizing gases [94] with the use of the conductometric method. These can be categorized into two types on grounds of their electronic structure [95]:

(a) Transition metal oxides (Fe_2O_3, NiO, Cr_2O_3, etc.)

(b) Non-transition metal oxides – i) Pre-transition (Al_2O_3, MgO)

ii) Post-transition (ZnO, SnO_2)

In transition metal oxides, the energy gap between a cation with d^n configuration and adjacent configurations (d^{n+1}, d^{n-1}) is quite small [96], which makes them more responsive to external changes. Non-transition metal oxides can be with d^0(TiO_2, V_2O_5, WO_3) or d^{10} configuration (ZnO, SnO_2) and are more extensive as compared to the transition metal oxides.

Semiconducting metal-oxide based sensing [97] is convenient, cost-effective and easy to undertake. The basic idea behind analyte gas detection is the reversibility of surface reaction between a metal oxide and analyte gas. The surface reaction involves adsorption of atmospheric oxygen to the metal oxide layer by electron capture from the bulk material. This influences the resistance of the semiconducting sensor. When the air (O_2) modified surface is exposed to analyte gas, it reacts with the surface adsorbed oxygen or oxide species, leading to a variation in the resistance thereby generating sensor response. Quantification of sensor signal helps to evaluate analyte gas concentrations. Fig. 38 [98] diagrammatically explains the working phenomena.

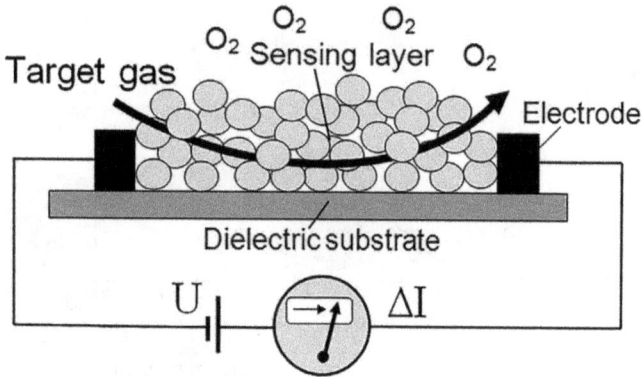

Fig. 38 *Diagrammatic explanation of working of metal-oxide based gas sensors. Reproduced here with the permission of Elsevier [98].*

Basically, two kinds of gaseous analyte behaviors have been encountered [99]:

1) those which reversibly adsorb and desorb generating reproducible results for all detections. Toxic gas CO (carbon monoxide) comes under this category.

2) the ones that undergo irreversible adsorption-desorption and introduces a modification in the sensor after initial detection. The SO_2 gas falls in this category.

The sensor construction primarily involves knowledge of two basic components (Fig. 39) [98]:

> ➤ sensing element – shows receptor function – undergoes surface reaction with the gas
> ➤ transducing element – measures the change in resistance after gas reception

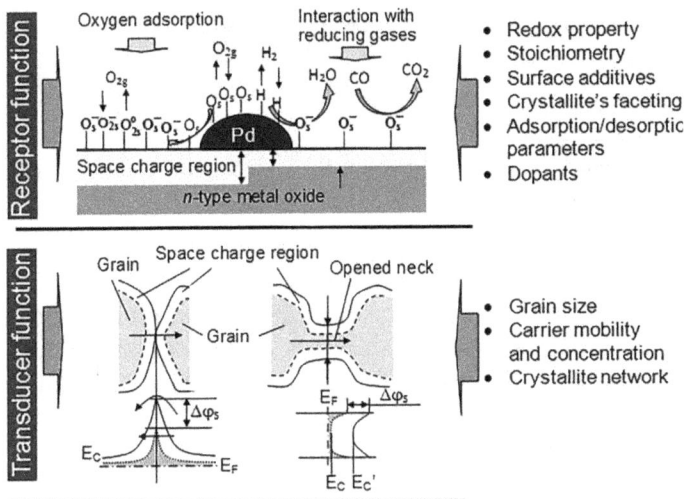

Fig. 39 *Properties of receptor and transducer function in semiconducting metal oxide sensors. Reproduced here with the permission of Elsevier [98].*

Variations in them give rise to a wide variety of sensors that can serve many applications.

In the following sections, three kinds of metal-oxide based sensors will be dealt with in somewhat detail.

i. Tin oxide (SnO₂) based sensors
ii. Titanium oxide (TiO₂) based sensors
iii. Tungsten oxide (WO₃) based sensors
iv. Zinc oxide (ZnO) based sensors

i. Tin-oxide (SnO₂) based sensors

Tin(IV) oxide is an example of non-stoichiometric [100] solids with inadequate oxygen atoms. Therefore, Sn(II) ions replace Sn(IV) ionic moieties to maintain the charge neutrality of the solid, thereby acting as electron-donor. Thus, SnO_2 is typically an n-type semiconductor. As mentioned earlier, the sensor works by adsorption of oxygen at raised temperature as the primary step, forming O_2^-, O^- or O^{2-}. In case of reducing gas acting as the analyte, electrons are released at the surface facilitating conduction and consequently

Materials Research Forum LLC
https://doi.org/10.21741/9781644901236

reduces the resistance of the surface layers. Conversely, oxidizing gases increase resistance [100].

The merits related to the use of this particular type include compact size, low operational voltage, high sensitivity, etc. The associated demerits are continual power outflow (approx. 0.5 W) due to sensor heating, instability with respect to certain temperatures and humid environments.

The selectivity ofsuch gas sensors can be controlled by doping of catalytic species like Pd, Cu, Ni, Pt [101], by varying the operating temperatures (25°-500°C), etc.

Berger et al. [102] have tried to explain the interaction of SO_2 and SnO_2 during its primary detection on the SnO_2 surface. On the basis of variations in acid-base conditions of SnO_2 sensing layers, followed by SO_2 exposure, and focusing on OH (hydroxyl) importance in such interactions have helped them in establishing the said mechanism (Fig. 40) [102].

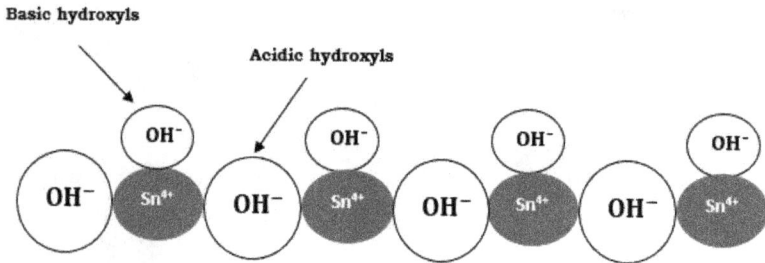

Fig. 40 *Characterization of surface hydroxyls in hydrated SnO₂ layers [102].*

Acquiring the primal understanding from Boehm et al. [103] regarding the types of OH groups in hydrated SnO_2 (Fig. 40) [102], the acidic OH sites of SnO_2 have been blocked with NH_3 and then carried out the reaction with SO_2. Following the same concept, basic OH sites were blocked by BF_3 before reacting with SO_2. The results revealed that acidic OH sites have no role to play in the SO_2/SnO_2 interaction in SO_2 primary detection; whereas it is the basic OH sites at which sulfate formation takes place during its first detection, altering the SnO_2 film for subsequent detections. Fig.41 (a) & (b) [102] signify the reaction of SnO_2 film with BF_3, followed by SO_2 exposure and the associated IR of the final product, respectively (Fig. 42) [102]. Fig. 43 [102] and Fig. 44 [102] depicts a similar reaction with $NH_3$and associated IR spectrum, respectively.

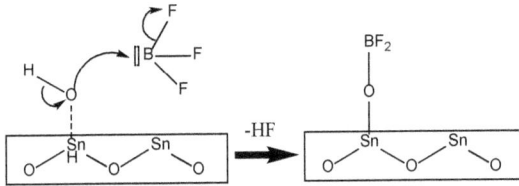

Fig. 41 *SnO₂ sites blocked with BF₃ and then reacted by SO₂ [102].*

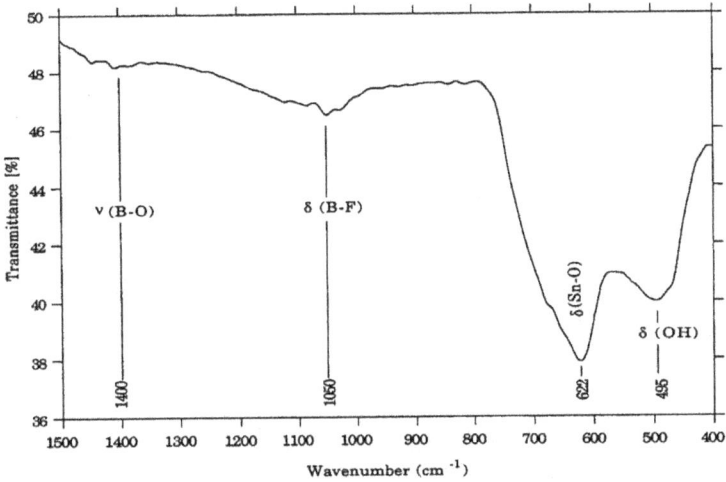

Fig. 42 *The infrared spectrum of SnO₂ reacted consecutively with BF₃ and SO₂. Reproduced here with the permission of Elsevier [102].*

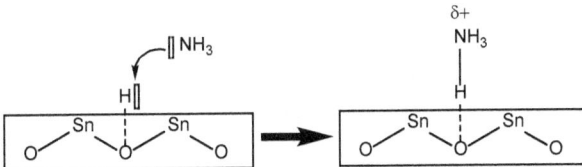

Fig. 43 *SnO₂ sites blocked with NH₃ and then reacted by SO₂ [102].*

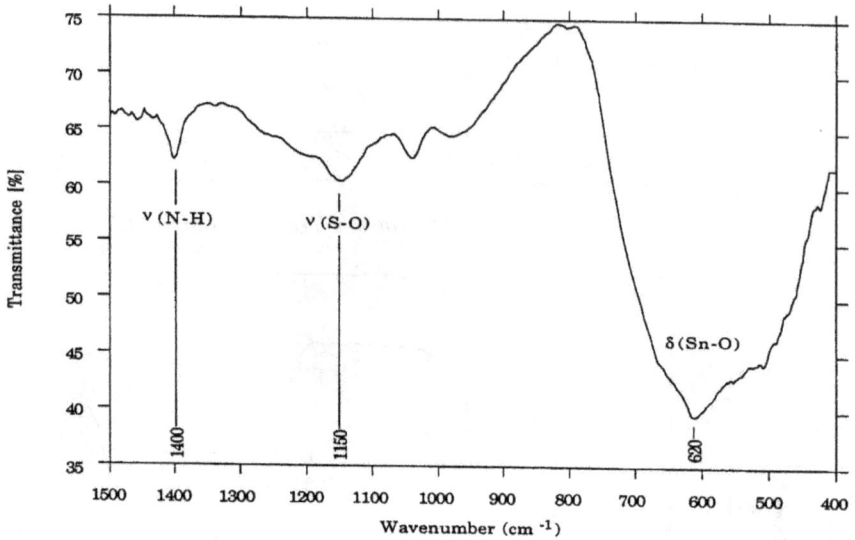

Fig. 44 *The infrared spectrum of SnO₂ reacted consecutively with NH₃ and SO₂.*
Reproduced here with the permission of Elsevier [102].

To enhance the sensor response attributes, doping is performed using various promoters. Das et al. [104] have investigated the effects of vanadium doping on SnO₂ sensing properties and received lower detection limits (5 ppm). Further improvement in sensor characteristics has been witnessed by Lee et al. [18], who have arrived at a dopant mixture – 5%wt MgO & 2%wt V₂O₅ that could afford high sensor response (44%) at 1.0 ppm SO₂ with decent repeatability. In their investigation, MgO was found to improvise the sensor response (maybe due to its SO₂ intake capacity), while V₂O₅ was thought to be responsible for better sensor reversibility. Sensor increased response and enhanced reversibility after doping is shown in Fig. 45 [18] and 46 [18] for 1.0 ppm SO₂ concentration respectively.

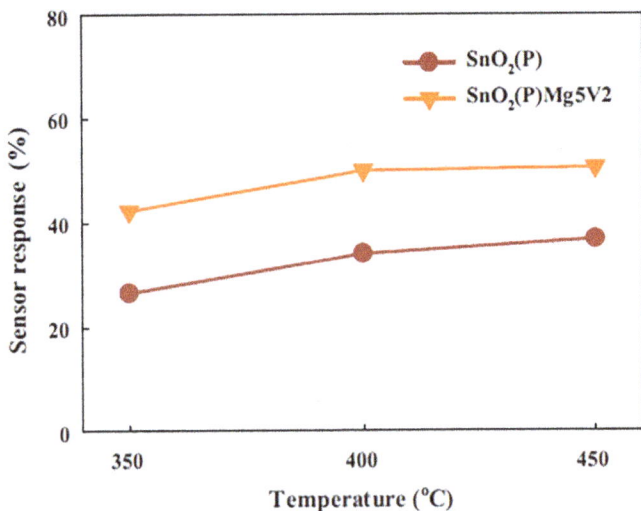

Fig. 45 *Responses of sensors SnO₂(P) and SnO₂(P)Mg₅V₂ at 1.0 ppm SO₂ at a temperature ranging from 350°-450 °C. Reproduced here with the permission of Elsevier [18].*

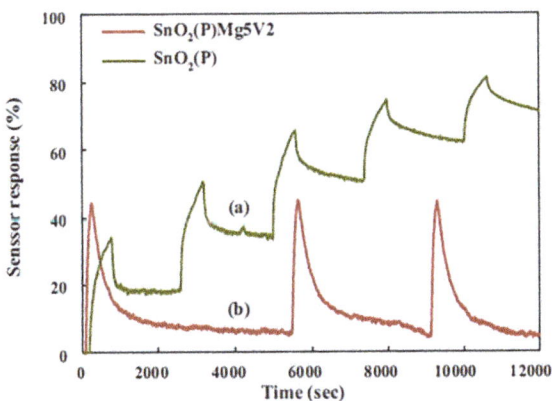

Fig. 46 *Reversibility of sensor responses of SnO₂(P) and SnO₂(P)Mg₅V₂ at 1.0 ppm SO₂ at 400 °C. Reproduced here with the permission of Elsevier [18].*

Tyagi et al. [105] have utilized metal oxide nanoclusters for customizing the SnO_2 thin film matrix to attain a low detection limit at a low operating temperature of 180°C. Among other metal oxide catalysts tested, NiO-nanostructures displayed the highest response (~56) at 180°C to 500 ppm SO_2 concentration in 80 s response time. Fig. 47 [105] illustrates the plot of the response signal of different hetero-structures (metal oxide/SnO_2) analyzed vs operating temperature.

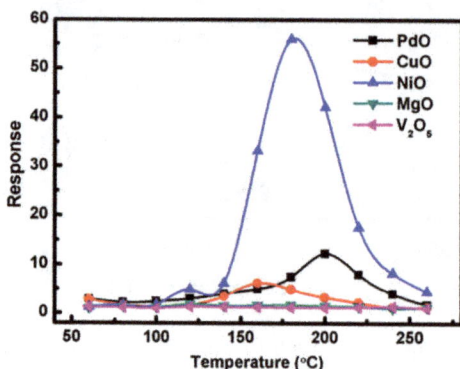

Fig. 47 *Sensors response to 500 ppm SO_2 at variable temperatures depicting the highest value corresponding to NiO. Reproduced here with the permission of Elsevier [105].*

The efficiency of SnO_2 can also be ameliorated with uniform deposition of noble metal over it. Research studies convey that noble metal catalytic activity has a role to play in chemisorption of molecular oxygen over SnO_2 surfaces [106-108] and hence may assist in the detection in two ways:

(a) by building up additional sites for gaseous uptake

(b) by lowering the activation energy requirements of the surface reactions [109].

Moreover, oxygen vacancies aid in the activation of oxygen molecules for reacting with analyte like gas molecules [110].

Liu and Liu [111] have extended this approach for the development of the AuNPs-amended SnO_2 gas sensor. The sensor yielded low detecting values ~500 ppb at low temperature (200°C), high selectivity and superb repeatability on account of oxygen vacancies. The introduction of oxygen vacancies in the AuNPs-SiO_{2-x} arrangement work by interceding the reaction of electrons, absorbing oxygen and hampering the charge carrier mobility, forming an electron depletion layer. Moreover, the Schottky contact (a rectifying contact between metal and SnO_2) allows electrons to flow from SnO_2 to Au metal thus creating a Schottky barrier. The extended depletion layer of SnO_2 nanocomposites facilitates high response upon SO_2 exposure. Fig. 48 [111] represents the mechanism of detection by the sensor.

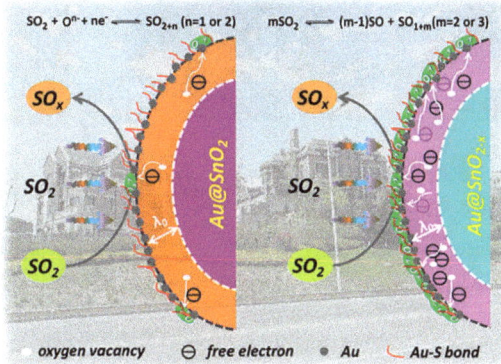

Fig. 48 Scheme of SO_2 Gas Sensing by AuNPs-SnO$_{2(-x)}$. Reproduced here with the permission of ACS [111].

A very recent innovation by Zhong et al. [112] dealing with tin oxide-based SO_2 sensors involves the growth of SnO_2 nanowires (SNWs) over a unique diatomite-based porous substrate. An illustrative sketch of the sensor is given in Fig. 49 [112].

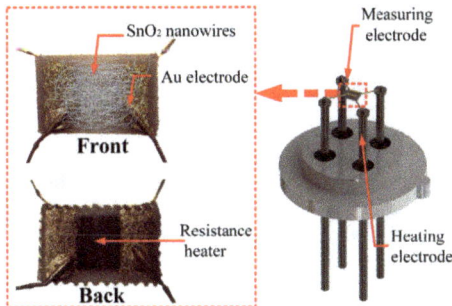

Fig. 49 Illustrative sketch of the typical sensor. Reproduced here with the permission of Elsevier [112].

The surface morphology of the substrate can help in controlling the length-to-diameter ratio of SNWs, an important feature concerning the enhanced gas-sensing performance. Fig. 50 [112] outlines the growth process of SNWs over the porous substrate.

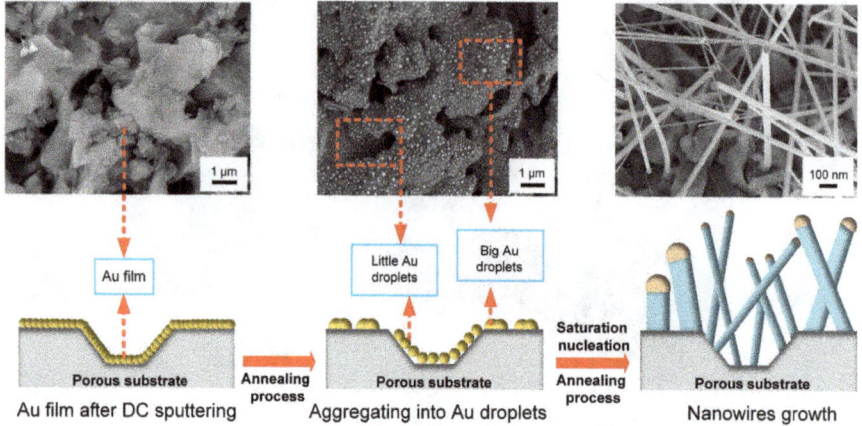

Fig. 50 *Schematic representation of the growth process of the SNWs developed on the diatomite-based porous substrate. Reproduced here with the permission of Elsevier [112].*

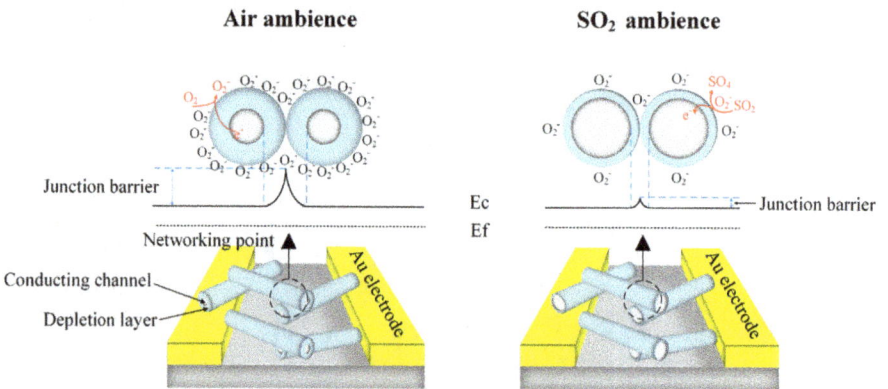

Fig. 51 *Demonstration of gas detection mechanism of SNWs [112].*

Materials Research Forum LLC
https://doi.org/10.21741/9781644901236

Large surface area and enormous wire-to-wire contacts of the SNWs render high resistance value to the sensor in ambient air (absence of analyte gas-SO_2). On SO_2 exposure, a reaction occurs between adsorbed oxygen moieties and SO_2 to free the confined e's back to the conduction band of SNWs, thus intensifying the response characteristics of the fabricated sensor. Fig. 51[112] explains the working principle of the sensor. The sensor can be used for sensing as low as 1.0 ppm giving maximum response value of ~33.5 for 50 ppm SO_2 at an optimal temperature of 85°C in a short response time of 3 s.

ii. Titanium oxide (TiO₂) based sensors

Besides the extensive applications of SnO_2 semiconducting films in toxic gas sensing, titanium oxide is all the better investigated. As it has been mentioned earlier, TiO_2 belongs to d^0 configuration non-transition metal oxides. Alterations in the titanium oxide sensing layers offer a wide potential for innovations to achieve high sensing characteristics. For example, graphene-based nano-assemblies are interesting candidates that offer high electronic conductivity, surface area, nominal noise and large stability at high temperatures [113, 114], being perfectly suitable for low-level detection of the target gas. A 2D arrangement of sp^2 hybridized carbon atoms is normally found in such assemblies. Its derivatives include graphene oxide (GO) and reduced graphene oxide (rGO) – a matter of much attraction due to ease of manufacturing and novel utilization [115].

Zhang et al. [116] have shared a method of preparing self-assembled titania/modified-graphene film layer-wise to sense ultralow levels of SO_2 (1.0 to 5.0 ppm) at room temperature. Fig. 52 [116] describes the layer-by-layer formation of TiO2/rGO nanocomposites.

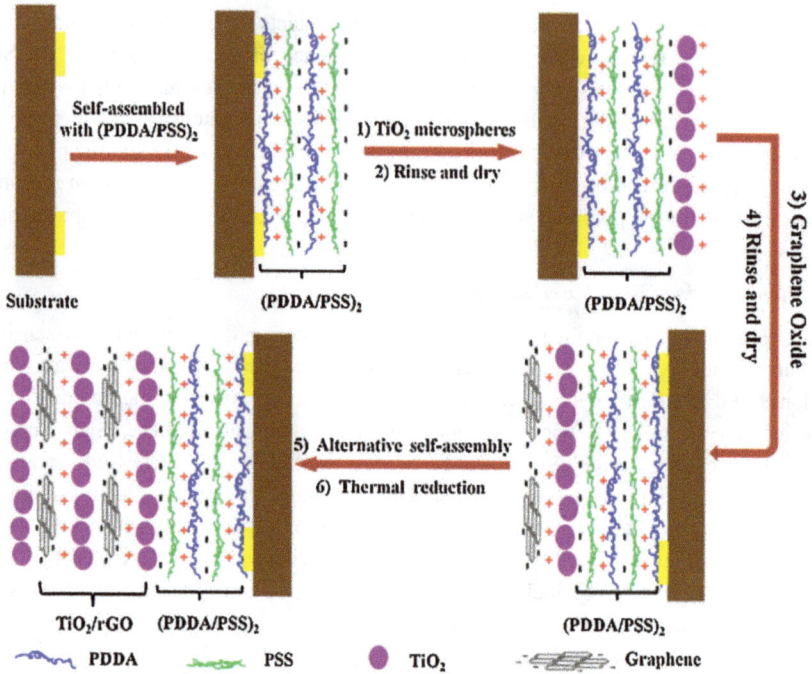

Fig. 52 *Layer-by-layer formation of TiO2/rGO nanocomposites. Reproduced here with the permission of Elsevier [116].*

The sensor working principle upon SO_2 exposure is similar as has been discussed earlier for SnO_2 sensors (n-type). The difference lies in the heterojunction created by the assembly of p-type rGO and n-type TiO_2, where the depletion layer is formed. This TiO_2/rGO hybrid structure behaves as an n-type semiconductor with respect to SO_2 in the surroundings. The dilution of the depletion layer in the presence of SO_2 is the cause of the sensor signal. Fig. 53 [116] explains the above phenomena pictorially.

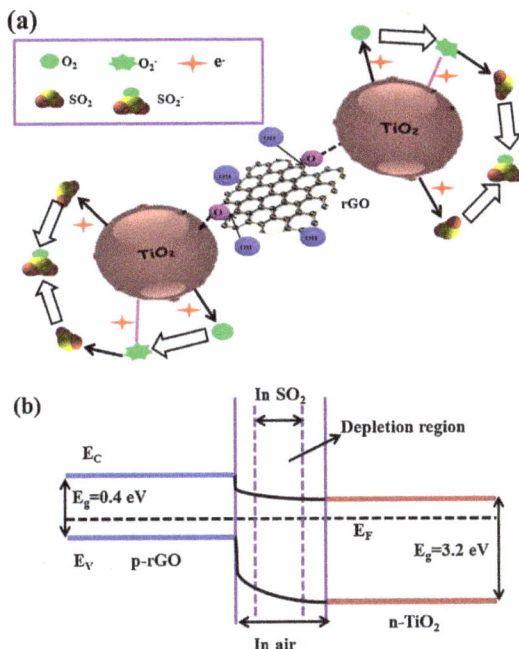

Fig. 53 *(a) Mechanism of SO₂ detection and (b) Energy band diagram of TiO₂/rGO nanocomposites. Reproduced here with the permission of Elsevier [116].*

In the course of developing sensors with unfamiliar detection techniques, Li et al. [117] have presented a distinctive and fascinating photo-electrochemical methodology employing Ba^{2+} doped TiO_2 for sulfur dioxide gas sensing. In this work, at first, the ITO electrode was tailored using amorphous titanium oxide hollow spheres forming photo-anodes bearing a photocurrent of 25.1 nA after Xe flash. The photocurrent slightly reduced upon Ba^{2+} ion doping. Before exposing it to SO_2, the SO_2 gas was procured in an aqueous medium by the oxidizing action of H_2O_2 followed by NaOH treatment. The resulting Na_2SO_4 (proportional to gaseous SO_2) on coming in contact with "Ba^{2+} doped TiO_2 photoanode" generated non-conducting $BaSO_4$ nanoparticles that eventually collapsed the photo-induced e^- pathway and photocurrent was substantially reduced. Fig. 54 [117] schematically depicts the photo-electrochemical sensing of SO_2 using the designed photo-anode. The corresponding sensor response is shown in Fig. 55 [117].

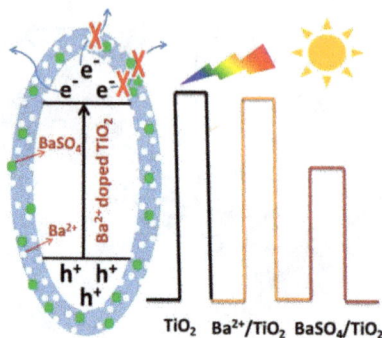

Fig. 54 *Photoelectrochemical sensing of SO₂ using the designed Ba²⁺ doped TiO₂ photoanode. Reproduced here with the permission of Elsevier [117].*

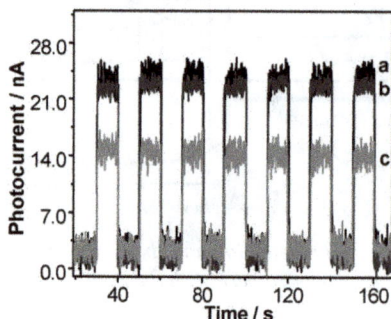

Fig. 55 *Photocurrent signal of (a)amorphous TiO₂/ITO, (b)Ba2+ doped TiO₂/ITO & (c)BaSO₄ nanoparticles doped TiO₂/ITO in 0.1M PBS (pH 7.0) containing 1.0 nM Na₂SO₄ voltage of 0.2 V with Xe lamp at an applied (150W). Reproduced here with the permission of Elsevier [117].*

The sensor displayed picomolar level sensitivity with a detection limit of 0.4 pM. The exceptional sensitivity of TiO_2 surfaces is attributed to Ba^{2+} doping that has deactivated the active sites phenomenally, thereby disallowing the photo-induced e⁻ to travel between the TiO_2 surface and ITO electrode.

iii. Tungsten oxide (WO₃) based sensors

Another type of d^0 configuration metal oxide that has been explored for gas sensing purposes is tungsten oxide. The WO_3 nanomaterials with varying morphologies have been developed and analyzed by Boudiba et al. [118] in the presence of low SO_2 concentrations ranging from 1-10 ppm at 200-300 °C. The electrical response of 10 μm thin film of WO_3 nanomaterials screen-printed over alumina surface has been recorded. The WO_3 nanoplates and nanowires provided a maximum response with the least time of response. The WO3 nanowires are nanoparticles displayed poor response at 260 °C in comparison to other materials, which deteriorated further at a lower temperature. Fig. 56 [118] displays the XRD pattern of morphologies of WO_3 nanopowders utilized for sensor fabrication and their TEM images.

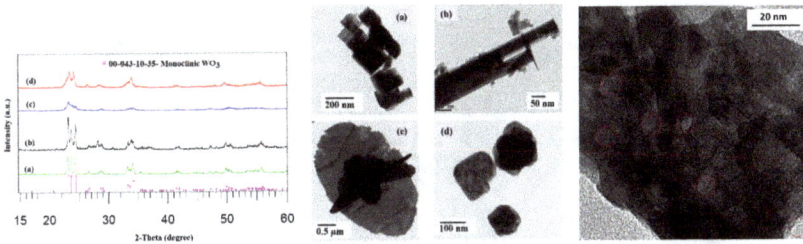

Fig. 56 *(1) XRD patterns and (2) TEM images, of WO_3 nanopowder (3) Porosity of sample(d) WO_3 nanoparticles depicted by HRTEM images; (a)-Nanoplates: 10 nm thick/100 nm lateral size, (b)Nanowires: 50 nm diameter/2 μm length, (c) Nanolamellae: 100 nm/2.5 μm, (d)Nanoparticles: 60 nm particle size. Reproduced here with the permission of Elsevier [118].*

iv. Zinc oxide (ZnO) based sensors

The ZnO based metal-oxide sensors have also been exploited for gaseous analytes [119-121]. These are n-type semiconductors that belong to the d^{10} configuration post-transition metal oxides category. Studies have revealed the novel features of ZnO that contribute to its gas sensing abilities; these include broad energy bandgap- 3.37eV, high e⁻ mobility, convenient method of preparation, exceptional stability with respect to temperature and chemical environment, etc. Research reports mention the use of various doping materials to further augment the sensing properties, e.g.: NiO (p-type) in conjunction with ZnO (n-type) has been made use of for toxic gas detection [122, 123].

NiO-ZnO nanodisks have recently been reported as electrode substances by Zhou et al. [124] for SO_2 identification at low concentration. The well-defined pattern of nanodisks

Materials Research Forum LLC
https://doi.org/10.21741/9781644901236

with 60±5 nm has been observed through surface morphological characterization techniques. Fig. 57 [124] showcases the Field emission scanning electron microscopy (FESEM) images along with the Energy dispersive X-ray spectroscopy (EDS) spectrum of NiO-ZnO heterostructures.

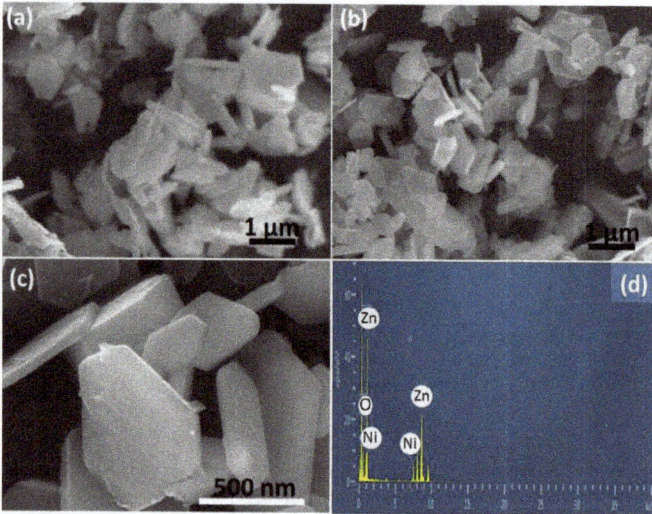

Fig. 57 *(a)-(c)FESEM images (b) EDS spectrum of NiO-ZnO nanodisks developed.
Reproduced here with the permission of Elsevier [124].*

The chemisorption of sulfur dioxide produced a change in the electrical resistance of the metal oxide semiconductor. The sulfur dioxide on metal oxide surface produced sulfur trioxide by oxygenated anionic species. Linearity in the gas response was observed in the 5-10 ppm range with the lowest detectable value as 3.0 ppm. This has been demonstrated in Fig. 58 (a) & (b) [124] for different SO₂ levels- in the range 5-800 ppm (curvilinear behavior) and 5-100 ppm (linear behavior), respectively. The sensor displayed excellent repeatability and stability at the working temperature.

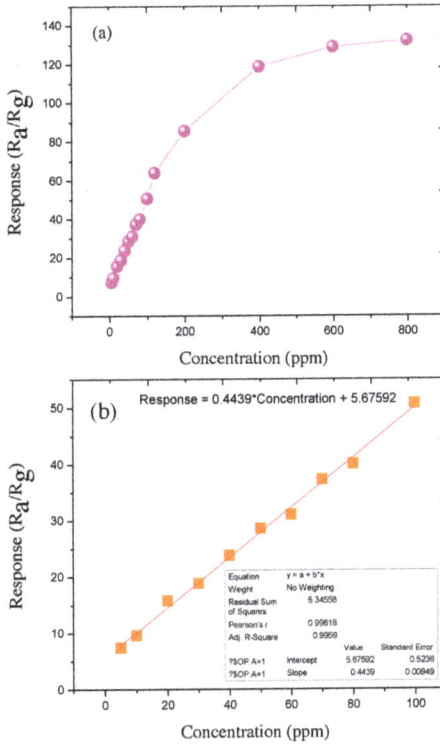

Fig. 58 *NiO-ZnO nanodisks based sensor response to varying SO$_2$ concentrations ranging from (a) 5–800ppm and (b) 5–100ppm at 240 °C. Reproduced here with the permission of Elsevier [124].*

(VI) Sensors based on photoacoustic detection

Photoacoustic spectroscopy (PAS) [125,126] is a technique that quantifies acoustic feedback produced on the interaction of incident electromagnetic radiation with a sample (maybe gas or condensed phase matter). The electromagnetic radiation when absorbed causes excitation of gaseous molecules to an upper electronic, vibrational or rotational quantum state. Non-radiative collisional relaxation to lower quantum states heats up the gas due to the energy being transferred to translation. Thus, the pressure is developed inside the photoacoustic cell containing the gaseous sample and leads to the generation of an acoustic wave detectable with the help of a microphone fitted in the instrument. The amplitude of the sound wave directly varies with the concentration of gas probed molecules. This technique is highly preferred for trace amount detection of a sample.

In pulsed-laser photoacoustic spectroscopy, the laser source is utilized for sample irradiation. This being highly monochromatic bears an advantage of selective excitation of molecular species at appropriate absorption wavelengths. Gondal and Dastageer [127] have achieved great sensitivity towards SO_2 detection by employing a UV-laser (266 nm). Fig.59 [127] describes the sensor design chosen for SO_2 sensing. The lowest limit of detection evaluated was 4.0 ppbv (parts per billion by volume).

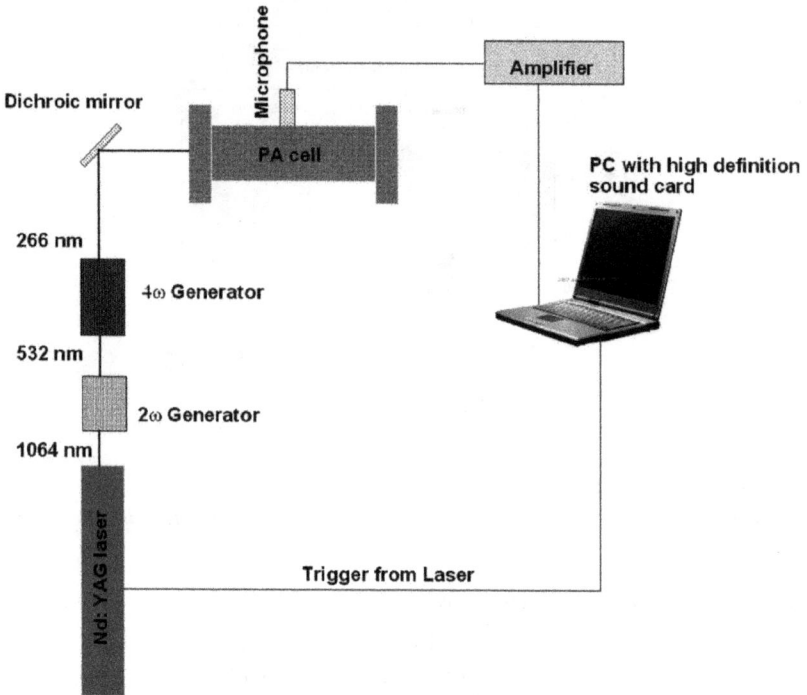

Fig. 59 *A sensor design was chosen for SO_2 detection. Reproduced here with the permission of Taylor & Francis [127].*

Materials Research Forum LLC
https://doi.org/10.21741/9781644901236

On the basis of advancement in transducing elements that produce the acoustic signal in response to pressure wave, Quartz-enhanced PAS has come into existence. This technique employs quartz tuning fork (QTF) that possesses high-quality factor > 10,000 as opposed to electric microphone in conventional PAS with a relatively low Q-factor~200. QTF is a piezoelectric item that resonates at ~2^{15} Hz in a vacuum and generates an electrical signal in response to the pressure waves developed inside the cell. With reference to its compact size, QTF enables the recognition of a very minute concentration of analytes. Waclawek et al.[128] have exploited this technique with the utilization of high heat load (HHL) packaged CW-DFB-QCL (continuous wave-distributed feedback-quartz cascade laser) source for detecting SO_2 at ppbv levels with minimum possible detectable concentration at 63ppbv. Fig. 60 [128] illustrates the sensor assembly. The sensor displayed linear performance as a function of SO_2 concentration (Fig. 61) [128]. Moreover, it was concluded with a futuristic approach to further lower the limit of detection approx. 8.2 times with the application of better QCL beam quality.

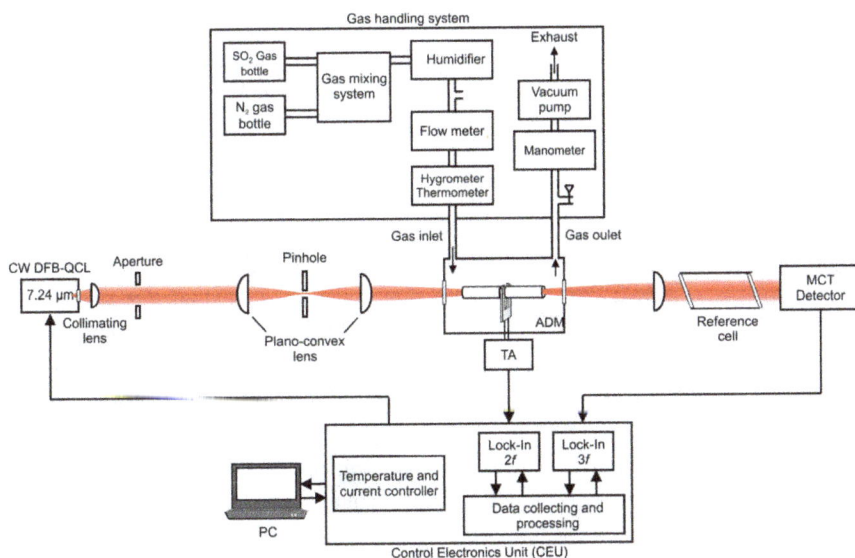

Fig. 60 *QEPAS-based SO₂ gas sensor using a 7.24 μm CW DFB-QCL. Reproduced here with the permission of Springer Nature [128].*

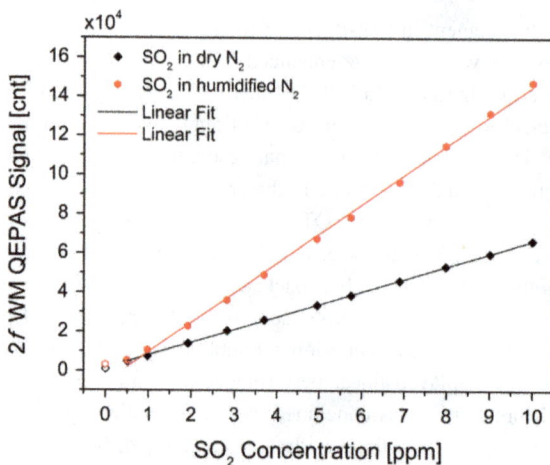

Fig. 61 *Linear performance of sensor as a function of SO₂ concentration (black: dry gas mixture, red: with 2.3 % absolute humidity moisturized gas mixture). Reproduced here with the permission of Springer Nature [128].*

5. Biosensing of SO₂

The biosensor is an analytical device that comprises two basic components, biological sensing element and a transducer. The biosensing moiety interacts with the analyte molecules to result in a biochemical change. Biomolecules such as antibodies, enzymes, receptor proteins, nucleic acids, whole-cell/tissue section [129]; and microorganisms can serve as the biosensing element. Transducer transforms the signal generated by the analyte-biosensing moiety interaction to a measurable response, like current, potential, absorption of light [130], etc. Depending on the type of response generated by the transducer, biosensors can be classified as electrochemical, optical or colorimetric.

In the development of a biosensor design, immobilization [131] serves as a pre-requisite. In this technique, the biological component is immobilized onto the sensor substrate, keeping in view that it is not very far from the transducer, for effective detection of the analyte. Moreover, the biosensing element should not be damaged in the process. Immobilization should also impart stability to the biological entity. It can be done using four methods, namely, adsorption, entrapment, covalent binding, and cross-linking.

Nowadays, the subject "biosensors" has become an attractive area for researchers. The reason being their low cost, high selectivity and sensitivity [129], easy transportation, mass production, short-time response, and high stability. These can be used in real-time analysis which helps in monitoring prompt changes [132] occurring in the surroundings.

Hart et al.[133], have developed an amperometric SO_2 biosensor employing screen-printed carbon electrodes (SPCEs). They have made use of the enzyme sulfite oxidase (SOD) and cytochrome c to be immobilized onto the SPCEs. This biosensor operates on a reaction scheme which firstly involves the dissolution of SO_2 (present in the environment) in the thin buffer layer coated on the surface of the biosensor. Here it is converted to SO_3^{2-}, which is then oxidized to sulfate by the enzyme (SOD) present at the active site. SOD itself gets reduced in the presence of Mo-cofactor and heme. The reduced SOD reacts with oxidized cytochrome c to generate reduced cytochrome c, which is re-oxidized at the SPCE to produce anodic current responsible for the analytical signal (Fig. 62) [133].

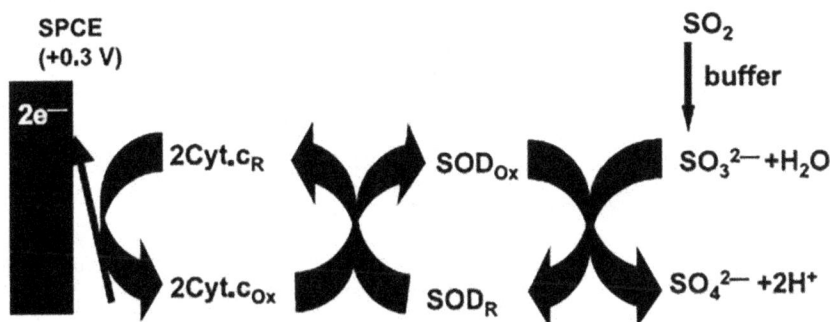

Fig. 62 *Scheme of reaction taking place as the biosensor works. Reproduced here with the permission of Elsevier [133].*

They designed two types of biosensors, viz. s-type and b-type, differing in the method of integration of enzyme and cytochrome c with SPCEs. Both the devices successfully detected SO_2 in the range 4 to 50 ppm producing linear responses in low response time (Fig. 63) [133].

A   **B**

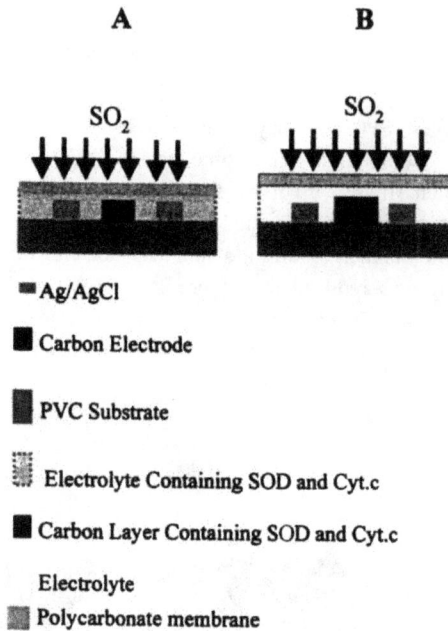

- Ag/AgCl

■ Carbon Electrode

▌ PVC Substrate

▒ Electrolyte Containing SOD and Cyt.c

■ Carbon Layer Containing SOD and Cyt.c

Electrolyte
▒ Polycarbonate membrane

***Fig**. 63 A schematic representation of (A) s-type biosensor and (B) b-type biosensor. Reproduced here with the permission of Elsevier [133].*

Conclusions

Alarming pollution levels pose a serious threat to human health. Sulfur dioxide is one of the major pollutants and contributes to the formation of suspended particulate matter responsible for impaired visibility and severe respiratory problems. Keeping in view the graveness of the issues, sensors have been developed to monitor the SO_2 emissions to enable the necessary steps that can be taken to counter these accordingly. This review summarises the sensors based on electrochemical sensing, optical (colorimetric and fluorescent methods), semiconducting metal oxide films incorporating various innovations to achieve selective and sensitive detection. The unconventional and pioneering techniques involved the use of ionic liquids which due to their high potential window, high gaseous solubility, high conductivity, were employed as electrolytes in amperometric/potentiometric sensing. In addition, high sensitivity to the trace amounts of gaseous analytes was attainable by the application of photoacoustic sensors. Each and every sensing technology offers great potential to be further explored for quick, selective and sensitive gas sensing so that these can be utilized in practical environment monitoring. However, much still need to be done to implement the sensing technologies in the remote area in a cost-effective manner.

Acknowledgments

The authors sincerely thank DST-SERB and DRDO, India for financial support (EMR/2016/005022 and ERIP/ER/DG-NSM/990116702/M/01/1645). The authors are thankful to the Principal, St. Stephen's College for the necessary infrastructure.

References

[1] E. Barea, C. Montoro, J.A.R. Navarro, Toxic gas removal – metal–organic frameworks for the capture and degradation of toxic gases and vapours, Chem. Soc. Rev. 43 (2014) 5419-5430. https://doi.org/10.1039/C3CS60475F

[2] S. Kumar, S. Chawla, M.C. Zou, Calixarenes based materials for gas sensing applications: a review, J. Incl. Phenom. Macrocycl. Chem. 88 (2017) 129-158. https://doi.org/10.1007/s10847-017-0728-2

[3] D. Fang, B. Chen, K. Hubacek, R. Ni, L. Chen, K. Feng, J. Lin, Clean air for some: Unintended spillover effects of regional air pollution policies, Sci. Adv. 5 (2019) eaav4707. https://doi.org/10.1126/sciadv.aav4707

[4] F.M. DaMatta, E. Rahn, P. Läderach, R. Ghini, J.C. Ramalho, Why could the coffee crop endure climate change and global warming to a greater extent than previously estimated?, Clim. Change 152 (2019) 167-178. https://doi.org/10.1007/s10584-018-2346-4

[5] R. Tricker, S. Tricker, Pollutants and contaminants, in: R. Tricker, S. Tricker (Eds.), Environmental requirements for electromechanical and electronic equipment, Newnes, Oxford, 1999, pp. 158-194.

[6] W. Bleam, Acid-base chemistry, in: W. Bleam (Ed.), Soil and environmental chemistry (Second edition), Academic Press, Waltham, MA,2017, pp. 253-331.

[7] A. Lazen, Emergency and continuous exposure limits for selected airborne contaminants, National Academy Press, Washington, D. C., 1984.

[8] D. Eatough, F. Caka, R. Farber, The conversion of SO_2 to sulfate in the atmosphere, Isr. J. Chem. 34 (1994) 301-314. https://doi.org/10.1002/ijch.199400034

[9] X. Pan, Sulfur oxides: sources, exposures and health effects, 1st ed., Elsevier, Amsterdam, Netherlands, 2011.

[10] B.G. Miller, The effect of coal usage on human health and the environment, in: B.G. Miller (Ed.), Clean coal engineering technology (Second edition), Butterworth-Heinemann, Burlington, MA, 2017, pp. 105-144

[11] S.J. Smith, J.V Aardenne, Z. Klimont, R.J. Andres, A. Volke, S. Delgado Arias, Anthropogenic sulfur dioxide emissions: 1850–2005, Atmos. Chem. Phys. 11 (2011) 1101-1116. https://doi.org/10.5194/acp-11-1101-2011

[12] D.A. Vallero, Fundamentals of air pollution, Academic press, London, 2014.

[13] N.R. Council, Air quality and stationary source emission control, National Academies Press, Washington, USA, 1975.

[14] X. Liu, S. Cheng, H. Liu, S. Hu, D. Zhang, H. Ning, A survey on gas sensing technology, Sensors (Basel) 12 (2012) 9635-9665. https://doi.org/10.3390/s120709635

[15] T. Anderson, F. Ren, S. Pearton, B.S. Kang, H.T. Wang, C.Y. Chang, J. Lin, Advances in hydrogen, carbon dioxide, and hydrocarbon gas sensor technology using GaN and ZnO-based devices, Sensors (Basel) 9 (2009) 4669-4694. https://doi.org/10.3390/s90604669

[16] H. Yoon, Current trends in sensors based on conducting polymer nanomaterials, Nanomaterials 3 (2013) 524-549. https://doi.org/10.3390/nano3030524

[17] P. Pasierb, M. Rekas, Solid-state potentiometric gas sensors—current status and future trends, J. Solid State Electr. 13 (2009) 3-25. https://doi.org/10.1007/s10008-008-0556-9

[18] S.C. Lee, B.W. Hwang, S.J. Lee, H.Y. Choi, S.Y. Kim, S.Y. Jung, D. Ragupathy, D.D. Lee, J.C. Kim, A novel tin oxide-based recoverable thick film SO_2 gas sensor promoted with magnesium and vanadium oxides, Sens. Actuator B-Chem. 160 (2011) 1328-1334. https://doi.org/10.1016/j.snb.2011.09.070

[19] W. Weppner, Solid-state electrochemical gas sensors, Sens. Actuator 12 (1987) 107-119. https://doi.org/10.1016/0250-6874(87)85010-2

[20] N. Yamazoe, N. Miura, Potentiometric gas sensors for oxidic gases, J. Electroceram. 2 (1998) 243-255. https://doi.org/10.1023/A:1009974506712

[21] M. Gauthier, A. Chamberland, Solid-state detectors for the potentiometric determination of gaseous oxides: I. Measurement in air, J. Electrochem. Soc. 124 (1977) 1579-1583. https://doi.org/10.1149/1.2133113

[22] B.-K. Min, S.-D. Choi, SO_2-sensing characteristics of Nasicon sensors with Na_2SO_4-

BaSO$_4$ auxiliary electrolytes, Sens. Actuator B-Chem. 93 (2003) 209-213. https://doi.org/10.1016/S0925-4005(03)00210-7

[23] T. Maruyama, Y. Saito, Y. Matsumoto, Y. Yano, Potentiometric sensor for sulfur oxides using NASICON as a solid electrolyte, Solid State Ion. 17 (1985) 281-286.

[24] J. Ross, J. Riseman, J. Krueger, Potentiometric gas sensing electrodes, Pure Appl. Chem. 36 (1973) 473-487. https://doi.org/10.1016/0167-2738(85)90072-4

[25] J.W. Severinghaus, A.F. Bradley, Electrodes for blood pO$_2$ and pCO$_2$ determination, J. Appl. Physiol. 13 (1958) 515-520. https://doi.org/10.1152/jappl.1958.13.3.515

[26] W. Worrell, Q. Liu, A new sulphur dioxide sensor using a novel two-phase solid-sulphate electrolyte, J. Electroanal. Chem. Interf. Electrochem. 168 (1984) 355-362. https://doi.org/10.1016/0368-1874(84)87109-9

[27] D.M. Pranitis, M.E. Meyerhoff, Sulfite-sensitive solvent/polymeric-membrane electrode based on bis(diethyldithiocarbamato)mercury(II), Anal. Chim. Acta 217 (1989) 123-133. https://doi.org/10.1016/S0003-2670(00)80392-X

[28] M.D. Mowery, R.S. Hutchins, P. Molina, M. Alajarín, A. Vidal, L.G. Bachas, Guanidinium-based potentiometric SO$_2$ gas sensor, Anal. Chem. 71 (1999) 201-204. https://doi.org/10.1021/ac980335n

[29] M. Madou, S. Morrison, Chemical sensing with solid state devices, Academic Press, London,1989.

[30] F. Lalère, J.B. Leriche, M. Courty, S. Boulineau, V. Viallet, C. Masquelier, V. Seznec, An all-solid state NASICON sodium battery operating at 200 °C, J. Power Sources 247 (2014) 975-980. https://doi.org/10.1016/j.jpowsour.2013.09.051

[31] J.B. Goodenough, H.Y.P. Hong, J.A. Kafalas, Fast Na$^+$-ion transport in skeleton structures, Mater. Res. 11 (1976) 203-220. https://doi.org/10.1016/0025-5408(76)90077-5

[32] C. Masquelier, L. Croguennec, Polyanionic (Phosphates, silicates, sulfates) frameworks as electrode materials for rechargeable Li (or Na) batteries, Chem. Rev. 113 (2013) 6552-6591. https://doi.org/10.1021/cr3001862

[33] T. Zhong, B. Quan, X. Liang, F. Liu, B. Wang, SO$_2$-sensing characteristics of NASICON sensors with ZnSnO$_3$ sensing electrode, Mater. Sci. Eng., B 151 (2008) 127-132. https://doi.org/10.1016/j.mseb.2008.05.010

[34] X. Liang, T. Zhong, B. Quan, B. Wang, H. Guan, Solid-state potentiometric SO$_2$ sensor combining NASICON with V$_2$O$_5$-doped TiO$_2$ electrode, Sens. Actuator B-

Materials Research Forum LLC
https://doi.org/10.21741/9781644901236

Chem. 134 (2008) 25-30. https://doi.org/10.1016/j.snb.2008.04.003

[35] Y. Shimizu, M. Okimoto, N. Souda, Solid-state SO_2 sensor using a sodium-ionic conductor and a metal–sulfide electrode, Int. J. Appl. Ceram. Tec. 3 (2006) 193-199. https://doi.org/10.1111/j.1744-7402.2006.02078.x

[36] H. Zhang, J. Yi, X. Jiang, Fast Response, Highly sensitive and selective mixed-potential H2 sensor based on (La, Sr)(Cr, Fe) O_3-δ perovskite sensing electrode, ACS Appl. Mater. Interfaces 9 (2017) 17218-17225. https://doi.org/10.1021/acsami.7b01901

[37] Y. Zhang, J. Zhang, J. Zhao, Z. Zhu, Q. Liu, Ag–$LaFeO_3$ fibers, spheres, and cages for ultrasensitive detection of formaldehyde at low operating temperatures, Phys. Chem. Chem. Phys. 19 (2017) 6973-6980. https://doi.org/10.1039/C6CP08283A

[38] W. Zhang, C. Xie, G. Zhang, J. Zhang, S. Zhang, D. Zeng, Porous $LaFeO_3/SnO_2$ nanocomposite film for CO_2 detection with high sensitivity, Mater. Chem. Phys. 186 (2017) 228-236. https://doi.org/10.1016/j.matchemphys.2016.10.048

[39] C. Ma, X. Hao, X. Yang, X. Liang, F. Liu, T. Liu, C. Yang, H. Zhu, G. Lu, Sub-ppb SO_2 gas sensor based on NASICON and $La_xSm_{1-x}FeO_3$ sensing electrode, Sens. Actuator B-Chem. 256 (2018) 648-655. https://doi.org/10.1016/j.snb.2017.09.194

[40] X. Liang, F. Liu, T. Zhong, B. Wang, B. Quan, G. Lu, Chlorine sensor combining NASICON with $CaMg_3(SiO_3)_4$-doped CdS electrode, Solid State Ion. 179 (2008) 1636-1640. https://doi.org/10.1016/j.ssi.2008.01.004

[41] X. Liang, Y. He, F. Liu, B. Wang, T. Zhong, B. Quan, G. Lu, Solid-state potentiometric H2S sensor combining NASICON with Pr_6O_{11}-doped SnO_2 electrode, Sens. Actuator B-Chem. 125 (2007) 544-549. https://doi.org/10.1016/j.snb.2007.02.050

[42] H. Zhang, T. Zhong, R. Sun, X. Liang, G. Lu, Sub-ppm H2S sensor based on NASICON and $CoCr_{2-x}Mn_xO_4$ sensing electrode, RSC Adv. 4 (2014) 55334-55340. https://doi.org/10.1039/C4RA07249A

[43] F. Liu, Y. Wang, B. Wang, X. Yang, Q. Wang, X. Liang, P. Sun, X. Chuai, Y. Wang, G. Lu, Stabilized zirconia-based mixed potential type sensors utilizing $MnNb_2O_6$ sensing electrode for detection of low-concentration SO_2, Sens. Actuator B-Chem. 238 (2017) 1024-1031. https://doi.org/10.1016/j.snb.2016.07.145

[44] V.I. Ogurtsov, K. Twomey, G. Herzog, Development of an integrated electrochemical sensing system to monitor port water quality using autonomous robotic fish, in: S. Hashmi, G.F. Batalha, C.J. Van Tyne, B. Yilbas (Eds.),

Comprehensive Materials Processing, Elsevier, Oxford, 2014, pp. 317-351.

[45] J.R. Stetter, J. Li, Amperometric gas sensors:Areview, Chem. Rev. 108 (2008) 352-366. https://doi.org/10.1016/0039-9140(93)80002-9

[46] C.Y. Chiou, T.C. Chou, Amperometric sulfur dioxide sensors using the gold deposited gas-diffusion electrode, Electroanalysis 8 (1996) 1179-1182. https://doi.org/10.1002/elan.1140081221

[47] D.R. Shankaran, S.S. Narayanan, Chemically modified sensor for amperometric determination of sulphur dioxide, Sens. Actuator B-Chem. 55 (1999) 191-194. https://doi.org/10.1016/S0925-4005(99)00053-2

[48] A. Hodgson, P. Jacquinot, P. Hauser, Electrochemical sensor for the detection of SO_2 in the low-ppb range, Anal. Chem. 71 (1999) 2831-2837. https://doi.org/10.1021/ac9812429

[49] A. Hodgson, P. Jacquinot, L. Jordan, P. Hauser, Amperometric gas sensors with detection limits in the low ppb range, Anal. Chim. Acta 393 (1999) 43-48. https://doi.org/10.1016/S0003-2670(99)00189-0

[50] N. Elgrishi, K.J. Rountree, B.D. McCarthy, E.S. Rountree, T.T. Eisenhart, J.L. Dempsey, A practical beginner's guide to cyclic voltammetry, J. Chem. Educ. 95 (2018) 197-206. https://doi.org/10.1021/acs.jchemed.7b00361

[51] P. Jasiński, A. Nowakowski, Simultaneous detection of sulphur dioxide and nitrogen dioxide by Nasicon sensor with platinum electrodes, Ionics 6 (2000) 230-234. https://doi.org/10.1007/BF02374071

[52] P. Jasiński, A. Nowakowski, H. Teterycz, K. Wisniewski, Thick film sensor based on NASICON for gas mixture detection, Ionics 5 (1999) 64-69. https://doi.org/10.1007/BF02375905

[53] G. Jasinski, P. Jasinski, B. Chachulski, A. Nowakowski, Electrocatalytic gas sensors based on Nasicon and Lisicon, Mater. Sci.-Poland 24 (2006) 261-268.. https://doi.org/10.1.1.485.268/BPW1-0021-0106

[54] E.L. Shoemaker, M.C. Vogt, F.J. Dudek, Cyclic voltammetry applied to an oxygen-ion-conducting solid electrolyte as an active electrocatalytic gas sensor, Solid State Ion. 92 (1996) 285-292. https://doi.org/10.1016/S0167-2738(96)00419-5

[55] G. Shi, M. Luo, J. Xue, Y. Xian, L. Jin, J.Y. Jin, The study of PVP/Pd/IrO_2 modified sensor for amperometric determination of sulfur dioxide, Talanta 55 (2001) 241-247. https://doi.org/10.1016/S0039-9140(01)00441-6

[56] H. Li, Q. Wang, J. Xu, W. Zhang, L. Jin, A novel nano-Au-assembled amperometric SO_2 gas sensor: preparation, characterization and sensing behavior, Sens. Actuator B-Chem. 87 (2002) 18-24. https://doi.org/10.1016/S0925-4005(02)00189-2

[57] J.J. Gooding, D.B. Hibbert, The application of alkanethiol self-assembled monolayers to enzyme electrodes, Trends Anal. Chem. 18 (1999) 525-533. https://doi.org/10.1016/S0165-9936(99)00133-8

[58] S. Ferretti, S. Paynter, D.A. Russell, K.E. Sapsford, D.J. Richardson, Self-assembled monolayers: a versatile tool for the formulation of bio-surfaces, Trends Anal. Chem. 19 (2000) 530-540. https://doi.org/10.1016/S0165-9936(00)00032-7

[59] D.R. Shankaran, N. Uehera, T. Kato, Determination of sulfur dioxide based on a silver dispersed functional self-assembled electrochemical sensor, Sens. Actuator B-Chem. 87 (2002) 442-447. https://doi.org/10.1016/S0925-4005(02)00304-0

[60] A. Strzelczyk, G. Jasiński, B. Chachulski, Construction and investigation of SO_2 amperometric sensor with a solid polymer electrolyte membrane, PhD Interdisciplinary J.3 (2013) 21-26.

[61] V.K. Rai, Temperature sensors and optical sensors, Appl. Phys. B 88 (2007) 297-303. https://doi.org/10.1007/s00340-007-2717-4

[62] H.N. Kim, W.X. Ren, J.S. Kim, J. Yoon, Fluorescent and colorimetric sensors for detection of lead, cadmium, and mercury ions, Chem. Soc. Rev. 41 (2012) 3210-3244. https://doi.org/10.1039/C1CS15245A

[63] A.V. Leontiev, D.M. Rudkevich, Revisiting noncovalent SO_2– amine chemistry: an indicator– displacement assay for colorimetric detection of SO_2, J. Am. Chem. Soc. 127 (2005) 14126-14127. https://doi.org/10.1021/ja053260v

[64] H. Yu, F. Dong, B. Li, F.S. Riehle, Co(II) triggered radical reaction between SO_2 and o-phenylenediamine for highly selective visual colorimetric detection of SO_2 gas and its derivatives, Sens. Actuator B-Chem. 299 (2019) 126983. https://doi.org/10.1016/j.snb.2019.126983

[65] S. Yin, P. Ruffin, Fiber optic sensors, in: M. Akay (Ed.) Wiley encyclopedia of biomedical engineering, Wiley-Interscience, Hoboken, N.J, 2006.

[66] O.S. Wolfbeis, A. Sharma, Fibre-optic fluorosensor for sulphur dioxide, Anal. Chim. Acta 208 (1988) 53-58. https://doi.org/10.1016/S0003-2670(00)80735-7

[67] L.K. Fraiji, D.M. Hayes, T. Werner, Static and dynamic fluorescence quenching experiments for the physical chemistry laboratory, J. Chem. Educ. 69 (1992) 424. https://doi.org/10.1021/ed069p424

Materials Research Forum LLC
https://doi.org/10.21741/9781644901236

[68] C. Huber, K. Fähnrich, C. Krause, T. Werner, Synthesis and characterization of new chloride-sensitive indicator dyes based on dynamic fluorescence quenching, J. Photochem. Photobiol. 128 (1999) 111-120. https://doi.org/10.1016/S1010-6030(99)00160-4

[69] J.R. Lakowicz, Principles of fluorescence spectroscopy, Springer Science & Business Media, New York, 2013.

[70] T.M. Razek, M.J. Miller, S.S. Hassan, M.A. Arnold, Optical sensor for sulfur dioxide based on fluorescence quenching, Talanta 50 (1999) 491-498. https://doi.org/10.1016/S0039-9140(99)00151-4

[71] A. Farooq, R. Al-Jowder, R. Narayanaswamy, M. Azzawi, P.J. Roche, D.E. Whitehead, Gas detection using quenching fluorescence of dye-immobilised silica nanoparticles, Sens. Actuator B-Chem. 183 (2013) 230-238. https://doi.org/10.1016/j.snb.2013.03.058

[72] M. Sun, H. Yu, K. Zhang, Y. Zhang, Y. Yan, D. Huang, S. Wang, Determination of gaseous sulfur dioxide and its derivatives via fluorescence enhancement based on cyanine dye functionalized carbon nanodots, Anal. Chem. 86 (2014) 9381-9385. https://doi.org/10.1021/ac503214v

[73] H. Li, H. Zhu, M. Sun, Y. Yan, K. Zhang, D. Huang, S. Wang, Manipulating the surface chemistry of quantum dots for sensitive ratiometric fluorescence detection of sulfur dioxide, Langmuir 31 (2015) 8667-8671. https://doi.org/10.1021/acs.langmuir.5b02340

[74] X. You, J. Wu, Y. Chi, Superhydrophobic silica aerogels encapsulated fluorescent perovskite quantum dots for reversible sensing of SO_2 in a 3D-printed gas cell, Anal. Chem. 91 (2019) 5058-5066. https://doi.org/10.1021/acs.analchem.8b05253

[75] S. Carn, V. Fioletov, C. McLinden, C. Li, N. Krotkov, A decade of global volcanic SO_2 emissions measured from space, Sci. Rep. 7, 44095, 2017. https://doi.org/10.1038/srep44095

[76] T. Bates, B. Lamb, A. Guenther, J. Dignon, R. Stoiber, Sulfur emissions to the atmosphere from natural sources, J. Atmos. Chem. 14 (1992) 315-337. https://doi.org/10.1007/BF00115242

[77] M. Tchalala, P. Bhatt, K. Chappanda, S. Tavares, K. Adil, Y. Belmabkhout, A. Shkurenko, A. Cadiau, N. Heymans, G. De Weireld, Fluorinated MOF platform for selective removal and sensing of SO_2 from flue gas and air, Nat. Commun. 10 (2019) 1328. https://doi.org/10.1038/s41467-019-09157-2

[78] P. Mahata, S.K. Mondal, D.K. Singha, P. Majee, Luminescent rare-earth-based MOFs as optical sensors, Dalton Trans. 46 (2017) 301-328. https://doi.org/10.1039/C6DT03419E

[79] M. Wang, L. Guo, D. Cao, Amino-functionalized luminescent metal–organic framework test paper for rapid and selective sensing of SO_2gas and its derivatives by luminescence turn-on effect, Anal.Chem. 90 (2018) 3608-3614. https://doi.org/10.1021/acs.analchem.8b00146

[80] P. Walden, Molecular weights and electrical conductivity of several fused salts, Bull. Acad. Imper. Sci.(St. Petersburg) 1800 (1914) 405-422.

[81] D. Wei, A. Ivaska, Applications of ionic liquids in electrochemical sensors, Anal. Chim. Acta 607 (2008) 126-135. https://doi.org/10.1016/j.aca.2007.12.011

[82] P. Wasserscheid, T. Welton, Ionic liquids in synthesis, John Wiley & Sons, Weinheim, Germany,2008.

[83] P.A. Suarez, V.M. Selbach, J.E. Dullius, S. Einloft, C.M. Piatnicki, D.S. Azambuja, R.F. de Souza, J. Dupont, Enlarged electrochemical window in dialkyl-imidazolium cation based room-temperature air and water-stable molten salts, Electrochim. Acta 42 (1997) 2533-2535. https://doi.org/10.1016/S0013-4686(96)00444-6

[84] X. Wang, J. Hao, Recent advances in ionic liquid-based electrochemical biosensors, Sci. Bull. 61 (2016) 1281-1295. https://doi.org/10.1007/s11434-016-1151-6

[85] Z. Wang, P. Lin, G.A. Baker, J. Stetter, X. Zeng, Ionic liquids as electrolytes for the development of a robust amperometric oxygen sensor, Anal. Chem. 83 (2011) 7066-7073. https://doi.org/10.1021/ac201235w

[86] A. Rehman, X. Zeng, Ionic liquids as green solvents and electrolytes for robust chemical sensor development, Acc. Chem. Res. 45 (2012) 1667-1677. https://doi.org/10.1021/ar200330v

[87] S. Zhang, Z.-l. Zhao, Y.X Wang, J. Gao, C.-b. Tan, Synthesis and photochromic properties of a series of spirooxazines [J], Chem. Res. Appl. 12 (2012). http://en.cnki.com.cn/Article_en/CJFDTotal-HXYJ201212003.htm

[88] D.S. Silvester, Recent advances in the use of ionic liquids for electrochemical sensing, Analyst 136 (2011) 4871-4882. https://doi.org/10.1039/C1AN15699C

[89] M.C. Buzzeo, C. Hardacre, R.G. Compton, Use of room temperature ionic liquids in gas sensor design, Anal. Chem. 76 (2004) 4583-4588. https://doi.org/10.1021/ac040042w

Materials Research Forum LLC
https://doi.org/10.21741/9781644901236

[90] X. Mu, Z. Wang, M. Guo, X. Zeng, A.J. Mason, Fabrication of a miniaturized room temperature ionic liquid gas sensor for human health and safety monitoring, 2012 IEEE Biomedical Circuits and Systems Conference (BioCAS), IEEE, 2012, pp. 140-143.

[91] L. Chen, Y. Zhang, S. Ren, D. Huang, C. Zhou, Y. Chi, G. Chen, An ionic liquid-mediated electrochemiluminescent sensor for the detection of sulfur dioxide at the ppb level, Analyst 138 (2013) 7006-7011. https://doi.org/10.1039/C3AN01407J

[92] S. Che, R. Dao, W. Zhang, X. Lv, H. Li, C. Wang, Designing an anion-functionalized fluorescent ionic liquid as an efficient and reversible turn-off sensor for detecting SO_2, Chem. Commun. 53 (2017) 3862-3865. https://doi.org/10.1039/C7CC00676D

[93] X. Xu, Z. Du, Y. Wang, X. Mao, L. Jiang, J. Yang, S. Hou, Electrochemical properties of a 2D-molybdenum disulfide–modified electrode and its application in SO_2 detection, J. Electroanal. Chem. 815 (2018) 220-224. https://doi.org/10.1016/j.jelechem.2018.03.020

[94] C. Wang, L. Yin, L. Zhang, D. Xiang, R. Gao, Metal oxide gas sensors: sensitivity and influencing factors, Sensors 10 (2010) 2088-2106. https://doi.org/10.3390/s100302088

[95] G. Korotcenkov, Metal oxides for solid-state gas sensors: What determines our choice?, Mater. Sci. Eng., B 139 (2007) 1-23. https://doi.org/10.1016/j.mseb.2007.01.044

[96] V.E. Henrich, P.A. Cox, The surface science of metal oxides, Cambridge university press, Cambridge, 1996.

[97] N. Barsan, U. Weimar, Fundamentals of metal oxide gas sensors, Proceedings IMCS-2012, 2012, pp. 618-621. https://doi.org/10.5162/IMCS2012/7.3.3

[98] G. Korotcenkov, B. Cho, Metal oxide composites in conductometric gas sensors: Achievements and challenges, Sens. Actuator B-Chem. 244 (2017) 182-210. https://doi.org/10.1016/j.snb.2016.12.117

[99] D. Girardin, F. Berger, A. Chambaudet, R. Planade, Modelling of SO_2 detection by tin dioxide gas sensors, Sens. Actuator B-Chem. 43 (1997) 147-153. https://doi.org/10.1016/S0925-4005(97)00149-4

[100] J. Watson, The tin oxide gas sensor and its applications, Sens. Actuator 5 (1984) 29-42. https://doi.org/10.1016/0250-6874(84)87004-3

[101] K. Ikohura, Tin oxide gas sensor for deoxydising gas, New Mater. and New

Progress in Electrochem. Tech. 1 (1981) 43-50.

[102] F. Berger, M. Fromm, A. Chambaudet, R. Planade, Tin dioxide-based gas sensors for SO_2 detection: a chemical interpretation of the increase in sensitivity obtained after a primary detection, Sens. Actuator B-Chem. 45 (1997) 175-181. https://doi.org/10.1016/S0925-4005(97)00284-0

[103] H. Boehm, Acidic and basic properties of hydroxylated metal oxide surfaces, Discuss. Faraday Soc. 52 (1971) 264-275. https://doi.org/10.1039/DF9715200264

[104] S. Das, S. Chakraborty, O. Parkash, D. Kumar, S. Bandyopadhyay, S. Samudrala, A. Sen, H.S. Maiti, Vanadium doped tin dioxide as a novel sulfur dioxide sensor, Talanta 75 (2008) 385-389. https://doi.org/10.1016/j.talanta.2007.11.010

[105] P. Tyagi, A. Sharma, M. Tomar, V. Gupta, Efficient detection of SO_2gas using SnO_2based sensor loaded with metal oxide catalysts, Procedia Eng. 87 (2014) 1075-1078. https://doi.org/10.1016/j.proeng.2014.11.349

[106] A.S.K. Hashmi, G.J. Hutchings, Gold catalysis, Angew. Chem. 45 (2006) 7896-7936. https://doi.org/10.1002/anie.200602454

[107] J.-H. Kim, P. Wu, H.W. Kim, S.S. Kim, Highly selective sensing of CO, C_6H_6, and C_7H_8 gases by catalytic functionalization with metal nanoparticles, ACS Appl. Mater. Interfaces 8 (2016) 7173-7183. https://doi.org/10.1021/acsami.6b01116

[108] H. Song, C. Li, Z. Lou, Z. Ye, L. Zhu, Effective formation of oxygen vacancies in black TiO_2 nanostructures with efficient solar-driven water splitting, ACS Sustain. Chem. Eng. 5 (2017) 8982-8987. https://doi.org/10.1021/acssuschemeng.7b01774

[109] C. Liu, Q. Kuang, Z. Xie, L. Zheng, The effect of noble metal (Au, Pd and Pt) nanoparticles on the gas sensing performance of SnO_2-based sensors: A case study on the {221} high-index faceted SnO_2 octahedra, CrystEngComm 17 (2015) 6308-6313. https://doi.org/10.1039/C5CE01162K

[110] M. Epifani, J.D. Prades, E. Comini, E. Pellicer, M. Avella, P. Siciliano, G. Faglia, A. Cirera, R. Scotti, F. Morazzoni, The role of surface oxygen vacancies in the NO_2sensing properties of SnO_2 nanocrystals, J. Phys. Chem. C 112 (2008) 19540-19546. https://doi.org/10.1021/jp804916g

[111] L. Liu, S. Liu, Oxygen vacancies as an efficient strategy for promotion of low concentration SO_2gas sensing: The case of au-modified SnO_2, ACS Sustain. Chem. Eng. 6 (2018) 13427-13434. https://doi.org/10.1021/acssuschemeng.8b03205

[112] X. Zhong, Y. Shen, S. Zhao, X. Chen, C. Han, D. Wei, P. Fang, D. Meng, SO_2

sensing properties of SnO_2 nanowires grown on a novel diatomite-based porous substrate, Ceram. Int. 45 (2019) 2556-2565. https://doi.org/10.1016/j.ceramint.2018.10.186

[113] K. Toda, R. Furue, S. Hayami, Recent progress in applications of graphene oxide for gas sensing: A review, Anal. Chim. Acta 878 (2015) 43-53. https://doi.org/10.1016/j.aca.2015.02.002

[114] D. Zhang, J. Liu, H. Chang, A. Liu, B. Xia, Characterization of a hybrid composite of SnO_2 nanocrystal-decorated reduced graphene oxide for ppm-level ethanol gas sensing application, RSC Adv. 5 (2015) 18666-18672. https://doi.org/10.1039/C4RA14611E

[115] D. Zhang, J. Liu, B. Xia, Layer-by-layer self-assembly of zinc oxide/graphene oxide hybrid toward ultrasensitive humidity sensing, IEEE Electron Device Lett. 37(7) (2016) 916-919. https://doi.org/10.1109/LED.2016.2565728

[116] D. Zhang, J. Liu, C. Jiang, P. Li, High-performance sulfur dioxide sensing properties of layer-by-layer self-assembled titania-modified graphene hybrid nanocomposite, Sens. Actuator B-Chem. 245 (2017) 560-567. https://doi.org/10.1016/j.snb.2017.01.200

[117] J. Li, W. Xie, R. Shao, X. Ju, H. Li, In situ Ba^{2+} exchange in amorphous TiO_2 hollow sphere for derived photoelectrochemical sensing of sulfur dioxide, Sens. Actuator B-Chem. 262 (2018) 282-288. https://doi.org/10.1016/j.snb.2018.01.199

[118] A. Boudiba, C. Zhang, C. Bittencourt, P. Umek, M.-G. Olivier, R. Snyders, M. Debliquy, SO_2 gas sensors based on WO_3 nanostructures with different morphologies, Procedia Eng. 47 (2012) 1033-1036. https://doi.org/10.1016/j.proeng.2012.09.326

[119] V.M. Latyshev, T.O. Berestok, A. Opanasyuk, A. Kornyushchenko, V.I. Perekrestov, Nanostructured ZnO films for potential use in LPG gas sensors, Solid State Sci. 67 (2017) 109-113. https://doi.org/10.1016/j.solidstatesciences.2017.02.010

[120] V. Galstyan, E. Comini, C. Baratto, G. Faglia, G. Sberveglieri, Nanostructured ZnO chemical gas sensors, Ceram. Int. 41 (2015) 14239-14244. https://doi.org/10.1016/j.ceramint.2015.07.052

[121] R. Kumar, O. Al-Dossary, G. Kumar, A. Umar, Zinc oxide nanostructures for NO_2 gas–sensor applications: A review, Nano-Micro Lett. 7 (2015) 97-120. https://doi.org/10.1007/s40820-014-0023-3

[122] L. Xu, R. Zheng, S. Liu, J. Song, J. Chen, B. Dong, H. Song, NiO@ZnO heterostructured nanotubes: coelectrospinning fabrication, characterization, and highly enhanced gas sensing properties, Inorg. Chem. 51 (2012) 7733-7740. https://doi.org/.10.1021/ic300749a

[123] K. Lokesh, G. Kavitha, E. Manikandan, G.K. Mani, K. Kaviyarasu, J.B.B. Rayappan, R. Ladchumananandasivam, J.S. Aanand, M. Jayachandran, M. Maaza, Effective ammonia detection using n-ZnO/p-NiO heterostructured nanofibers, IEEE Sens. J. 16 (2016) 2477-2483. https://doi.org/10.1109/JSEN.2016.2517085

[124] Q. Zhou, W. Zeng, W. Chen, L. Xu, R. Kumar, A. Umar, High sensitive and low-concentration sulfur dioxide (SO_2) gas sensor application of heterostructure NiO-ZnO nanodisks, Sens. Actuator B-Chem. 298 (2019) 126870. https://doi.org/10.1016/j.snb.2019.126870

[125] F.J. Harren, J. Mandon, S.M. Cristescu, Photoacoustic spectroscopy in trace gas monitoring, in: R.A. Meyers (Ed.) Encyclopedia of Analytical Chemistry: Applications, Theory and Instrumentation, John Wiley & Sons Inc, New York, 2020.

[126] R.W. Jones, J. McClelland, 6 - Fourier transform infrared photoacoustic spectroscopy of ageing composites, in: R. Martin (Ed.), Ageing of Composites, Woodhead Publishing, Cambridge, England,2008, pp. 160-185.

[127] M.A. Gondal, M.A. Dastageer, High-sensitivity detection of hazardous SO_2 using 266 nm UV laser, J. Environ. Sci. Heal. A 43 (2008) 1126-1131. https://doi.org/10.1080/10934520802171543

[128] J. Waclawek, R. Lewicki, H. Moser, M. Brandstetter, F. Tittel, B. Lendl, Quartz-enhanced photoacoustic spectroscopy-based sensor system for sulfur dioxide detection using a CW DFB-QCL, Appl. Phys. B 117 (2014) 113-120. https://doi.org/10.1007%2Fs00340-014-5809-y

[129] Y.B. Hahn, R. Ahmad, N. Tripathy, Chemical and biological sensors based on metal oxide nanostructures, Chem. Commun. 48 (2012) 10369-10385. https://doi.org/10.1039/C2CC34706G

[130] Y. Lei, W. Chen, A. Mulchandani, Microbial biosensors, Anal. Chim. Acta 568 (2006) 200-210. https://doi.org/10.1016/j.aca.2005.11.065

[131] I. Karube, Y. Nomura, Y. Arikawa, Biosensors for environmental control, Trends Anal. Chem. 14 (1995) 295-299. https://doi.org/10.1016/0165-9936(95)97055-6

[132] G. Rocchitta, A. Spanu, S. Babudieri, G. Latte, G. Madeddu, G. Galleri, S. Nuvoli,

P. Bagella, M.I. Demartis, V. Fiore, Enzyme biosensors for biomedical applications: Strategies for safeguarding analytical performances in biological fluids, Sensors 16 (2016) 780. https://doi.org/10.3390/s16060780

[133] J.P. Hart, A.K. Abass, D. Cowell, Development of disposable amperometric sulfur dioxide biosensors based on screen printed electrodes, Biosens. Bioelectron. 17 (2002) 389-394. https://doi.org/10.1016/S0956-5663(01)00308-6

About the authors

Loveleen Kaur Gulati

The author received a BSc (Hons.) in Chemistry from St. Stephen's College, University of Delhi (India) in 2013, and an MSc in Chemistry from the University of Delhi in 2015. She is working as a research fellow under the supervision of Dr. Satish Kumar at St. Stephen's College. The research aimed at the development of photochromic receptors for toxic analyte sensing.

Gurleen Kaur Gulati

The author received a BSc (Hons.) in Chemistry from St. Stephen's College, University of Delhi (India) in 2013, and an MSc in Chemistry from the University of Delhi in 2015. She is working as a research fellow under the supervision of Dr. Satish Kumar at St. Stephen's College. The research involved the synthesis of macrocyclic receptors for the detection of toxic ions.

Satish Kumar

The author is an assistant professor in the Department of Chemistry at St. Stephen's College, University of Delhi. During the last 18 years, Dr. Satish Kumar has worked in interdisciplinary areas covering theoretical chemistry, nanochemistry, and application of principles of molecular recognition to design molecular receptors. Dr. Satish Kumar has published several research papers related to the development of receptors for neutral molecules, anions, and cations. He has 16 years of teaching experience.

www.ingramcontent.com/pod-product-compliance
Lightning Source LLC
Chambersburg PA
CBHW071510210326
41597CB00018B/2713